AF553300

ENERGY SCIENCE

ENERGY SCIENCE

Dr. Ashok Kumar

RANDOM PUBLICATIONS

NEW DELHI - 110 002 (INDIA)

Energy Science

ISBN 978-93-51117-45-2

Published in 2015 in India by

RANDOM PUBLICATIONS

4376-A/4B, Gali Murari Lal, Ansari Road
New Delhi-110 002
Phone: +9111-43580356, 23289044
E-mail: randomexports@gmail.com; sales@randompublications.com; info@randompublications.com

Reprinted 2025

Type Setting by: Friends Media, Delhi-110089
Digitally Printed at: Replika Press Pvt. Ltd.

Preface

In physics, energy is a property of objects, transferable among them via fundamental interactions, which can be converted in form but not created or destroyed. The joule is the SI unit of energy, based on the amount transferred to an object by the mechanical work of moving it 1 metre against a force of 1 newton. Work and heat are two categories of processes or mechanisms that can transfer a given amount of energy. The second law of thermodynamics limits the amount of work that can be performed by energy that is obtained via a heating process—some energy is always lost as waste heat. The maximum amount that can go into work is called the available energy. Systems such as machines and living things often require available energy, not just any energy. Mechanical and other forms of energy can be transformed in the other direction into thermal energy without such limitations. There are many forms of energy, but all these types must meet certain conditions such as being convertible to other kinds of energy, obeying conservation of energy, and causing a proportional change in mass in objects that possess it. Common energy forms include the kinetic energy of a moving object, the radiant energy carried by light and other electromagnetic radiation, the potential energy stored by virtue of the position of an object in a force field such as a gravitational, electric or magnetic field, and the thermal energy comprising the microscopic kinetic and potential energies of the disordered motions of the particles making up matter. Some specific forms of potential energy include elastic energy due to the stretching or deformation of solid objects and chemical energy such as is released when a fuel burns. Any object that has mass when stationary, such as a piece of ordinary matter, is said to have rest mass, or an equivalent amount of energy whose form is called rest energy, though this isn't immediately apparent in everyday phenomena described by classical physics. According to mass–energy equivalence, all forms of energy exhibit mass. For example, adding 25 kilowatt-hours (90 megajoules) of energy to an object in the form of heat increases its mass by 1 microgram; if you had a sensitive enough mass balance or scale, this mass increase could be measured. Our Sun transforms nuclear potential energy to other forms of energy; its total mass does not decrease due to that in itself, but its mass does decrease when the energy escapes out to its surroundings, largely as radiant energy.

Although any energy in any single form can be transformed into another form, the law of conservation of energy states that the total energy of a system can only change if energy is transferred into or out of the system. This means that it is impossible to create or destroy energy. The total energy of a system can be calculated by adding up all forms of energy in the system. Examples of energy transfer and transformation include generating or making use of electric energy, performing chemical reactions, or lifting an object. Lifting against gravity performs work on the object and stores gravitational potential energy; if it falls, gravity does work on the object which transforms the potential energy to the kinetic energy associated with its speed. More broadly, living organisms require available energy to stay alive; humans get such energy from food along with the oxygen needed to metabolize it. Civilisation requires a supply of energy to function; energy resources such as fossil fuels are a vital topic in economics and politics. Earth's climate and ecosystem are driven by the radiant energy Earth receives from the sun, and are sensitive to changes in the amount received. The word "energy" is also used outside of physics in many ways, which can lead to ambiguity and inconsistency. The vernacular terminology is not consistent with technical terminology. For example, while energy is always conserved (in the sense that the total energy does not change despite energy transformations), energy can be converted into a form, e.g., thermal energy, that cannot be utilized to perform work. When one talks about "conserving energy by driving less", one talks about conserving fossil fuels and preventing useful energy from being lost as heat. This usage of "conserve" differs from that of the law of conservation of energy.

Covering renewable, fossil fuel, and nuclear energy sources, this book relates the science behind these sources to the environmental and socioeconomic issues which surround their use to provide a balanced, objective overview. The book will appeal to students, teachers and a broad range of energy professionals.

I thank all members of my team who have helped in the preparation of the book. My special thanks go to "Random Publications" who have published the book.

— Dr. Ashok Kumar

Contents

Chapter 1

Introduction

History of the Energy System

In the Beginning: Pre-Industrialization

The muscle power of human beings and animals was the first application of energy by humans and the food chain was the energy system in use. Humans have long "designed" energy systems with the goal of producing the most work possible with the least amount of human effort to generate the energy.

Pre-Industrial society depended primarily on muscle power and biomass for their energy needs. Biomass consisted primarily of wood or peat and its energy delivery had a low efficiency. Amory Lovins, an expert on energy, states, "Most of the energy generated by wood or peat went up in the chimneys rather than into the room or cooking pot of pre-industrial societies."

Animal power in the form of horse mills, wind power in the form of windmills, and water power with the use of a water wheel were major energy sources harnessed until the 19th century; especially for "industrial uses." Wood and charcoal were the main fuels for cooking, heating, and other domestic uses, but coal and oil were available as well. "In the Middle East crude oils have been known for millennia from natural seepage and pools, but they were used only rarely as fuels, and more frequently as protective coatings." Coal has its origin in "the lithification of peats produced by accumulations of dead plant matter in wetlands. Difference in original vegetation and, more importantly, in magnitudes of durations of transforming temperatures and pressures, have produced a large variety of coals." As early as the 13th century, coal pits were mined and coal energy was used specifically for the forcing and smelting of metals. In the 1600's, England experienced an energy crisis due to a shortage of wood and began using coal as a substitute fuel source for domestic purposes. Even in the 1700's, wood was the major fuel source in colonial America.

The Industrial Revolution

The quest for more powerful energy sources was propelled by the inventions and discoveries of the Industrial Revolution. As sophisticated mechanical inventions were made, a large reliable and seemingly inexhaustible source of energy became necessary for industrial uses, and transportation. The need for large quantities of accessible, dependable, and transportable energy encouraged the exploration of energy sources. The inventions of the Industrial Revolution provided the equipment to further mine or drill the already visible deposits of coal and oil.

Steam power was developed in the 1600's in conjunction with coal mining to help pump water out of the mines. It had been known since ancient times that heat could be used to produce steam, which could then do mechanical work. However, it was only in the late eighteenth century that commercially successful steam engines were invented. The first commercially successful steam engine was invented by Thomas Savery (1650-1715), an English military engineer. In 1712, this engine was refined by Thomas Newcomen (1663-1729), another Englishman. The Newcomen engine was widely used in Britain and Europe throughout the eighteenth century, but had very low energy efficiency.

A greatly improved steam engine was designed and built in 1763 by James Watt who was asked to repair a Newcomen engine. Watt built and then sold or rented his engines to mining companies, charging them for the "power" in the rate of work the engine produced. Today, the unit for power is called a Watt.

The sun was also studied as an energy source in the 18th century. In 1767, the first solar thermal collector was developed by the Swiss scientist Horace de Saussure. Solar thermal power was used in the American west as an energy source for cooking until oil and natural gas became a more reliable way to generate energy. For simple cooking solar energy was absorbed by black cast iron pots. Solar thermal collectors were also used in the form of hot boxes to cook food.

In 1839, Alexandre Becquerel discovered that an electric current could be generated when certain elements were exposed to light. The scientific explanation of this phenomenon by Albert Einstein, called photoelectricity (light-induced electricity), came much later in 1905. Photoelectricity is the basis of the photovoltaic cells, now used to convert light into electricity. Despite the century and a half since it discovery, photovoltaic means of generating electricity have not been developed with enough vigor for it to become a major source of electricity. This is because the material technology for photovoltaic panels developed slowly. As coal and other fossil fuels were easier to use, and available in plenty, not much effort has gone into photovoltaic research.

Until the early 1800's our understanding of the science of energy was not well developed. The theory at that time was the caloric theory, which said that heat is a substance called "caloric" that flowed from hotter to colder bodies. In the 1840's the English physicist James Prescott Joule did a long series of experiments that showed that heat is a form of energy. Joule found the relationship between a unit of mechanical energy and a unit of heat. This helped Joule finalize what chemists and natural philosophers had come to believe—that the total energy in the universe is constant, although energy is continuously changing forms.

The study and invention of the heat engine and steam power established and confirmed the Laws of Thermodynamics. From 1840-1880, Joule, Lord Kelvin, and James Clark Maxwell in England; Sadi Carnot and Rudolf Clausius in France; and Ludwig Boltzmann in Austria formulated a theory of heat engines, laying the foundations of Thermodynamics, literally the science of "motion from heat." (Thermo=heat and dynamics=motion).

In 1820, the advances in mechanical and materials engineering made the railroad the most efficient and fastest means of transportation. Coal and wood were used as the primary fuel source for the steam engine. The locomotive also changed society's perception of travel and transportation.

Wind energy was developed on a large scale in the United States as an energy source for farms and railroad stations, using tall windmills to pump water from underground wells. There were specific design developments that made these windmills more efficient, although they still generated relatively little power. The height of these windmills helped to ensure they caught the wind and a tailfin generally kept the fan facing the wind.

Another result of the Industrial Revolution was an energy distribution infrastructure built into cities that promoted domestic convenience. As early as 1816, natural gas was piped into cities for domestic uses such as cooking, home illumination, and street lighting. The steam engine was used to pump water into homes and sewage away from homes. The city was undergirded with networks that usually began with water pipes and gas lines and gradually expanded to include sewers, electrical conduits, and telephone lines.

In 1859, when petroleum was drilled in Titusville, Pennsylvania, an apparently plentiful energy source began to replace coal. Oil was distilled into kerosene (referred to as coal oil) and used as a lamp oil. It replaced dwindling supplies of whale oil used for lamps. There were many reasons oil became a more desirable fuel source than coal: it was easy to obtain and transport; it emitted less particulate pollution than coal; it replaced scarce whale oil as a fuel for lamps; and coal had become an unreliable fuel source because of the labour issues surrounding the mining of coal. Miners were striking for safer work environments and more money, which affected the amount of coal available to the consumer.

But the most significant use of crude oil was as the liquid fuel for the internal combustion engine, designed in 1861 by Nikolaus August Otto. The internal combustion engine became one of the most influential inventions of the Industrial Revolution. Although this engine is low in efficiency, it could produce enough work to move a large metal vehicle far distances. The fuel of the internal combustion engine was also easier to use than, for example, shoveling coal into a furnace to power a locomotive. This was the beginning of the use of liquid fuel to advance transportation.

In 1879, Thomas Edison invented the incandescent light bulb — a major step in the human use of storable energy leading eventually to large-scale electrification. Electricity is similar to a liquid fuel in that it can be transported easily (although not efficiently) from one place to another. One of Edison's goals was to make electricity affordable for all homes. Edison began with the distribution of electricity through a direct current (DC). This meant that electrons would flow one way through a wire to bring electricity to a home; however, a good portion of the energy was lost as the electrons moved through the wire. This loss of energy using direct current to move electricity meant that power plants had to be built close to the homes the plant serviced and was eventually considered impractical.

Nikola Tesla, an inventor employed by Edison, discovered that electrons would alternate or travel back and forth on a wire and travel longer distances with less energy loss. This was called alternating current (AC) and had an advantage because AC could be more easily generated. Edison had so much money invested in his DC power plants that he discredited Tesla's alternate current as dangerous — thus beginning a "war of the currents." Tesla eventually joined forces with George Westinghouse and began developing power plants using alternating current (AC).

In the late 19th and early 20th centuries the steam turbine, using coal as a fuel, was developed as a cheap power source that generated electricity. In 1882, the first functional steam turbine was designed by Charles Parsons, an English engineer. He used the high pressure of steam to hit the blades of a rotor. The principle of the turbine was a major step toward today's production of electricity. In 1893, Westinghouse demonstrated a "universal system" of generation and distribution at a Chicago exposition. The universal system meant that power or energy could be used in a variety of ways at many different voltages. Westinghouse, using Tesla's invention of the transformer and the electric motor, as well as steam turbines, transformed Niagara Falls into one of the first hydroelectric plants in the world.

In 1910, Henry Ford opened the 60-acre Highland Park automotive plant with a moving assembly line. This was the beginning of an eventually enormous use for fossil fuels. Fossil fuels were used not only to propel the automobiles that were made at the plant, but also to generate electric power for the automotive plant.

Energy technologies developed rapidly during the 20th century. Although the current version for solar thermal collectors was designed in 1908, they were not developed well enough for mass distribution. In the 1920's, 30's, and 40's, there was large-scale construction and development of hydroelectric plants/dams to support increasing population in the Southwest. In 1938, Otto Hahn and Fritz Strassman demonstrated nuclear fission and within four years (1942), Oak Ridge, Tennessee, was chosen as the site for the first functional nuclear reactor plant, and for the preparation of uranium and plutonium used to the create the atomic bomb at Los Alamos. The first nuclear chain reactor was demonstrated at the University of Chicago in December 1942. In July 1945, the testing of the first atomic bomb at Alamogordo, New Mexico, demonstrated the technology used to release nuclear energy on a large scale. In 1957, the first commercial nuclear power plant opened in Shippingport, Pennsylvania.

The first large scale use of photovoltaic (PV) solar energy in conjunction with satellite technology developed in the 1950's. The United States Vanguard I was the first PV-powered satellite. By the early part of the 20th century, crude oil and its products had become an indispensable part of the industrial economy. James Young had patented a process in England in 1850 to distill oil from coal and shale. Oil refining is not just about gasoline. The distilled chemicals from crude oil have many purposes — for example, petroleum is used for plastics manufacturing. Young's process of fractal distillation forms the basis of the world's oil refining industry. While a large amount of oil occurs in many parts of the world, the largest stores are located in the regions governed by the Arab countries. The Oil and Petroleum Economic Cartel (OPEC) is the economic coalition of these countries that control the flow of that oil. In the 1970's, OPEC placed an embargo on their oil sales. This "energy crisis" brought energy scarcity to the consciousness of all nations — and especially the U.S., with its higher dependence on imported oil. This crisis began to generate interest in the exploration of renewable energy sources for large-scale generation of electricity and other energy needs.

World Energy Consumption

World energy consumption refers to the total energy used by all of human civilization. Typically measured per year, it involves all energy harnessed from every energy source applied towards humanity's endeavours across every industrial and technological sector, across every country. Being the power source metric of civilization, World Energy Consumption has deep implications for humanity's social-economic-political sphere. Institutions such as the International Energy Agency (IEA), the U.S. Energy Information Administration (EIA), and the European Environment Agency record and publish energy data periodically. Improved data and understanding of World Energy Consumption may reveal systemic trends and patterns, which could help frame current energy issues and encourage movement towards collectively useful solutions.

Fossil energy use increased most in 2000-2008. In October 2012 the IEA noted that coal accounted for half the increased energy use of the prior decade, growing faster than all renewable energy sources. The 1979 Three Mile Island accident and 1986 Chernobyl disaster, along with high construction costs, ended the rapid growth of global nuclear power capacity.

Energy use in terawatt-hours				
	Fossil	***Nuclear***	***Renewable***	***Total***
1990	83,374	6,113	13,082	102,569
2000	94,493	7,857	15,337	117,687
2008	117,076	8,283	18,492	143,851
Change 2000-2008	+23.9%	+5.4%	+20.6%	+22.2%

1 terawatt-hour (TWh) = 1 billion kilowatt-hours (kWh) = 10^{12} watt-hours

Trends

The energy consumption growth in the G20 slowed down to 2% in 2011, after the strong increase of 2010. The economic crisis is largely responsible for this slow growth. For several years now, the world energy demand is characterized by the bullish Chinese and Indian markets, while developed countries struggle with stagnant economies, high oil prices, resulting in stable or decreasing energy consumption.

According to IEA data from 1990 to 2008, the average energy use per person increased 10% while world population increased 27%. Regional energy use also grew from 1990 to 2008: the Middle East increased by 170%, China by 146%, India by 91%, Africa by 70%, Latin America by 66%, the USA by 20%, the EU-27 block by 7%, and world overall grew by 39%.

In 2008, total worldwide energy consumption was 474 exajoules (132,000 TWh). This is equivalent to an average power use of 15 terawatts (2.0×10^{10} hp). The annual potential for renewable energy is:

- solar energy 1,575 EJ (438,000 TWh),
- wind power 640 EJ (180,000 TWh),
- geothermal energy 5,000 EJ (1,400,000 TWh),
- biomass 276 EJ (77,000 TWh),
- hydropower 50 EJ (14,000 TWh) and
- ocean energy 1 EJ (280 TWh).

Energy consumption in the G20 increased by more than 5% in 2010 after a slight decline of 2009. In 2009, world energy consumption decreased for the first time in 30 years, by –1.1%—equivalent to 130 megatonnes (130,000,000 long tons; 140,000,000 short tons) of oil—as a result of the financial and economic crisis, which reduced world GDP by 0.6% in 2009.

This evolution is the result of two contrasting trends: Energy consumption growth remained vigorous in several developing countries, specifically in Asia (+4%). Conversely, in OECD, consumption was severely cut by 4.7% in 2009 and was thus almost down to its 2000 levels. In North America, Europe and the CIS, consumptions shrank by 4.5%, 5% and 8.5% respectively due to the slowdown in economic activity. China became the world's largest energy consumer (18% of the total) since its consumption surged by 8% during 2009 (up from 4% in 2008). Oil remained the largest energy source (33%) despite the fact that its share has been decreasing over time. Coal posted a growing role in the world's energy consumption: in 2009, it accounted for 27% of the total.

Most energy is used in the country of origin, since it is cheaper to transport final products than raw materials. In 2008 the share export of the total energy production by fuel was: oil 50% (1,952/3,941 Mt), gas 25% (800/3,149 bcm), hard coal 14% (793/5,845 Mt) and electricity

1% (269/20,181 TWh). Most of the world's high energy resources are from the conversion of the sun's rays to other energy forms after being incident upon the planet. Some of that energy has been preserved as fossil energy, some is directly or indirectly usable; for example, via solar PV/thermal, wind, hydro- or wave power. The total solar irradiance is measured by satellite to be roughly 1,361 watts (1.825 hp) per square meter, though it fluctuates by about 6.9% during the year due to the Earth's varying distance from the sun. This value, after multiplication by the cross-sectional area intercepted by the Earth, is the total rate of solar energy received by the planet; about half, 89 petawatts (1.19×10^{14} hp), reaches the Earth's surface.

The estimates of remaining non-renewable worldwide energy resources vary, with the remaining fossil fuels totaling an estimated 0.4 YJ (1 YJ = 10^{24}J) and the available nuclear fuel such as uranium exceeding 2.5 YJ. Fossil fuels range from 0.6 to 3 YJ if estimates of reserves of methane clathrates are accurate and become technically extractable. The total power flux from the sun intercepting the Earth is 5.5 YJ/yr, though not all of this is available for human consumption. The IEA estimates for the world to meet global energy demand for the two decades from 2015 to 2035 it will require investment of $48 trillion and "credible policy frameworks."

According to IEA (2012) the goal of limiting warming to 2 °C is becoming more difficult and costly with each year that passes. If action is not taken before 2017, CO_2 emissions would be locked-in by energy infrastructure existing in 2017. Fossil fuels are dominant in the global energy mix, supported by $523 billion subsidies in 2011, up almost 30% on 2010 and six times more than subsidies to renewables.

Greenhouse Effect

The greenhouse effect is a process by which thermal radiation from a planetary surface is absorbed by atmospheric greenhouse gases, and is re-radiated in all directions. Since part of this re-radiation is back towards the surface and the lower atmosphere, it results in an elevation of the average surface temperature above what it would be in the absence of the gases.

Solar radiation at the frequencies of visible light largely passes through the atmosphere to warm the planetary surface, which then emits this energy at the lower frequencies of infrared thermal radiation. Infrared radiation is absorbed by greenhouse gases, which in turn re-radiate much of the energy to the surface and lower atmosphere. The mechanism is named after the effect of solar radiation passing through glass and warming a greenhouse, but the way it retains heat is fundamentally different as a greenhouse works by reducing airflow, isolating the warm air inside the structure so that heat is not lost by convection. If an ideal thermally conductive blackbody were the same distance from the Sun as the Earth is, it would have a temperature of about 5.3 °C. However, since the Earth reflects about 30% of the incoming sunlight, this idealized planet's effective temperature (the temperature of a blackbody that would emit the same amount of radiation) would be about –18 °C. The surface temperature of this hypothetical planet is 33 °C below Earth's actual surface temperature of approximately 14 °C. The mechanism that produces this difference between the actual surface temperature and the effective temperature is due to the atmosphere and is known as the greenhouse effect.

Earth's natural greenhouse effect makes life as we know it possible. However, human activities, primarily the burning of fossil fuels and clearing of forests, have intensified the natural greenhouse effect, causing global warming.

History

The existence of the greenhouse effect was argued for by Joseph Fourier in 1824. The argument and the evidence was further strengthened by Claude Pouillet in 1827 and 1838, and reasoned from experimental observations by John Tyndall in 1859, and more fully quantified by Svante Arrhenius in 1896. In 1917 Alexander Graham Bell wrote "[The unchecked burning of fossil fuels] would have a sort of greenhouse effect", and "The net result is the greenhouse becomes a sort of hot-house." Bell went on to also advocate for the use of alternate energy sources, such as solar energy.

Mechanism

The Earth receives energy from the Sun in the form UV, visible, and near IR radiation, most of which passes through the atmosphere without being absorbed. Of the total amount of energy available at the top of the atmosphere (TOA), about 50% is absorbed at the Earth's surface. Because it is warm, the surface radiates far IR thermal radiation that consists of wavelengths that are predominantly much longer than the wavelengths that were absorbed (the overlap between the incident solar spectrum and the terrestrial thermal spectrum is small enough to be neglected for most purposes). Most of this thermal radiation is absorbed by the atmosphere and re-radiated both upwards and downwards; that radiated downwards is absorbed by the Earth's surface. This trapping of long-wavelength thermal radiation leads to a higher equilibrium temperature than if the atmosphere were absent.

This highly simplified picture of the basic mechanism needs to be qualified in a number of ways, none of which affect the fundamental process.

- The incoming radiation from the Sun is mostly in the form of visible light and nearby wavelengths, largely in the range 0.2–4 μm, corresponding to the Sun's radiative temperature of 6,000 K. Almost half the radiation is in the form of "visible" light, which our eyes are adapted to use.
- About 50% of the Sun's energy is absorbed at the Earth's surface and the rest is reflected or absorbed by the atmosphere. The reflection of light back into space—largely by clouds—does not much affect the basic mechanism; this light, effectively, is lost to the system.
- The absorbed energy warms the surface. Simple presentations of the greenhouse effect, such as the idealized greenhouse model, show this heat being lost as thermal radiation. The reality is more complex: the atmosphere near the surface is largely opaque to thermal radiation (with important exceptions for "window" bands), and most heat loss from the surface is by sensible heat and latent heat transport. Radiative energy losses become increasingly important higher in the atmosphere largely because of the decreasing concentration of water vapor, an important greenhouse gas. It is more realistic to think of the greenhouse effect as applying to a "surface" in the mid-troposphere, which is effectively coupled to the surface by a lapse rate.
- The simple picture assumes a steady state. In the real world there is the diurnal cycle as well as seasonal cycles and weather. Solar heating only applies during daytime. During the night, the atmosphere cools somewhat, but not greatly, because its emissivity is low, and during the day the atmosphere warms. Diurnal temperature changes decrease with height in the atmosphere.
- Within the region where radiative effects are important the description given by the idealized greenhouse model becomes realistic: The surface of the Earth, warmed to a temperature around 255 K, radiates long-wavelength, infrared heat in the range 4–100 μm. At these wavelengths, greenhouse gases that were largely transparent to incoming solar radiation are more absorbent. Each layer of atmosphere with greenhouses gases absorbs some of the heat being radiated upwards from lower

layers. It re-radiates in all directions, both upwards and downwards; in equilibrium (by definition) the same amount as it has absorbed. This results in more warmth below. Increasing the concentration of the gases increases the amount of absorption and re-radiation, and thereby further warms the layers and ultimately the surface below.

- Greenhouse gases—including most diatomic gases with two different atoms (such as carbon monoxide, CO) and all gases with three or more atoms—are able to absorb and emit infrared radiation. Though more than 99% of the dry atmosphere is IR transparent (because the main constituents—N_2, O_2, and Ar—are not able to directly absorb or emit infrared radiation), intermolecular collisions cause the energy absorbed and emitted by the greenhouse gases to be shared with the other, non-IR-active, gases.

Greenhouse Gases

By their percentage contribution to the greenhouse effect on Earth the four major gases are:

- water vapor, 36–70%
- carbon dioxide, 9–26%
- methane, 4–9%
- ozone, 3–7%

The major non-gas contributor to the Earth's greenhouse effect, clouds, also absorb and emit infrared radiation and thus have an effect on radiative properties of the atmosphere.

Role in Climate Change

Strengthening of the greenhouse effect through human activities is known as the enhanced (or anthropogenic) greenhouse effect. This increase in radiative forcing from human activity is attributable mainly to increased atmospheric carbon dioxide levels. According to the latest Assessment Report from the Intergovernmental Panel on Climate Change, "*most of the observed increase in globally averaged temperatures since the mid-20th century is very likely due to the observed increase in anthropogenic greenhouse gas concentrations*".

CO_2 is produced by fossil fuel burning and other activities such as cement production and tropical deforestation. Measurements of CO_2 from the Mauna Loa observatory show that concentrations have increased from about 313 ppm in 1960 to about 389 ppm in 2010. It reached the 400ppm milestone on May 9, 2013. The current observed amount of CO_2 exceeds the geological record maxima (~300 ppm) from ice core data. The effect of combustion-produced carbon dioxide on the global climate, a special case of the greenhouse effect first described in 1896 by Svante Arrhenius, has also been called the Calendar effect.

Over the past 800,000 years, ice core data shows that carbon dioxide has varied from values as low as 180 parts per million (ppm) to the pre-industrial level of 270ppm. Paleoclimatologists consider variations in carbon dioxide concentration to be a fundamental factor influencing climate variations over this time scale.

Real Greenhouses

The "greenhouse effect" of the atmosphere is named by analogy to greenhouses which get warmer in sunlight, but the mechanism by which the atmosphere retains heat is different. A greenhouse works primarily by allowing sunlight to warm surfaces inside the structure, but then preventing absorbed heat from leaving the structure through convection, i.e. sensible heat transport. The "greenhouse effect" heats the Earth because greenhouse gases absorb outgoing radiative energy, heating the atmosphere which then emits radiative energy with some of it going back towards the Earth.

A greenhouse is built of any material that passes sunlight, usually glass, or plastic. It mainly heats up because the Sun warms the ground inside, which then warms the air in the greenhouse. The air continues to heat because it is confined within the greenhouse, unlike the environment outside the greenhouse where warm air near the surface rises and mixes with cooler air aloft. This can be demonstrated by opening a small window near the roof of a greenhouse: the temperature will drop considerably. It has also been demonstrated experimentally (R. W. Wood, 1909) that a "greenhouse" with a cover of rock salt (which is transparent to infra red) heats up an enclosure similarly to one with a glass cover. Thus greenhouses work primarily by preventing convective cooling.

In contrast, the greenhouse effect heats the Earth because rather than retaining (sensible) heat by physically preventing movement of the air, greenhouse gases act to warm the Earth by re-radiating some of the energy back towards the surface. This process may exist in real greenhouses, but is comparatively unimportant there.

Bodies Other than Earth

In the Solar System, Mars, Venus, and the moon Titan also exhibit greenhouse effects; that on Venus is particularly large, due to its atmosphere, which consists mainly of dense carbon dioxide. Titan has an anti-greenhouse effect, in that its atmosphere absorbs solar radiation but is relatively transparent to infrared radiation. Pluto also exhibits behaviour superficially similar to the anti-greenhouse effect.

A runaway greenhouse effect occurs if positive feedbacks lead to the evaporation of all greenhouse gases into the atmosphere. A runaway greenhouse effect involving carbon dioxide and water vapor is thought to have occurred on Venus.

Newton (Unit)

The newton (symbol: N) is the International System of Units (SI) derived unit of force. It is named after Isaac Newton in recognition of his work on classical mechanics, specifically Newton's second law of motion.

In 1946, Conférence Générale des Poids et Mesures (CGPM) resolution 2 standardized the unit of force in the MKS system of units to be the amount needed to accelerate 1 kilogram of mass at the rate of 1 metre per second squared. The 9th CGPM, held in 1948, then adopted the name "newton" for this unit in resolution 7. This name honours the English physicist and mathematician Isaac Newton, who laid the foundations for most of classical mechanics. The newton thus became the standard unit of force in le Système International d'Unités (SI), or International System of Units.

Newton's second law of motion states that $F = ma$, where F is the force applied, m is the mass of the object receiving the force, and a is the acceleration of the object. The newton is therefore:

$$1\ \mathrm{N} = 1\ \mathrm{kg \cdot m / s^2}$$

where the following symbols are used for the units:

N: newton

kg: kilogram

m: metre

s: second.

In dimensional analysis:

$$\mathsf{F} = \frac{\mathsf{ML}}{\mathsf{T}^2}$$

where

F: force

M: mass

L: length

T: time.

This SI unit is named after Isaac Newton. As with every International System of Units (SI) unit whose name is derived from the proper name of a person, the first letter of its symbol is upper case (N). However, when an SI unit is spelled out in English, it should always begin with a lower case letter (*newton*), except in a situation where *any* word in that position would be capitalized, such as at the beginning of a sentence or in capitalized material such as a title. Note that "degree Celsius" conforms to this rule because the "d" is lowercase.— Based on *The International System of Units*, section 5.2.

Examples

- 1 N is the force of Earth's gravity on a mass of about 102 g = ($^1/_{9.81}$ kg).
- On Earth's surface, a mass of 1 kg exerts a force of approximately 9.81 N [down] (or 1.0 kilogram-force; 1 kgf = 9.80665 N by definition). The approximation of 1 kgf corresponding to 10 N (1 decanewton or daN) is sometimes used as an approximation in everyday life and in engineering.
- The force of Earth's gravity on (= the weight of) a human being with a mass of 70 kg is approximately 686 N.
- The dot product of force and distance is mechanical work. Thus, in SI units, a force of 1 N exerted over a distance of 1 m is 1 N.m of work. The Work-Energy Theorem states that the work done on a body is equal to the change in energy of the body. 1 N.m = 1 J (joule), the SI unit of energy.
- It is common to see forces expressed in kilonewtons or kN, where 1 kN = 1,000 N.

Common Use of Kilonewtons in Construction

Kilonewtons are often used for stating safety holding values of fasteners, anchors, and more in the building industry. They are also often used in the specifications for rock climbing equipment. The safe working loads in both tension and shear measurements can be stated in kilonewtons. Injection moulding machines, used to manufacture plastic parts, are classed by kilonewton (i.e., the amount of clamping force they apply to the mould).

On the Earth's surface, 1 kN is about 101.97162 kilogram-force of load, so multiplying the kilonewton value by 100 (i.e. using a slightly conservative and easier to calculate value) is a good approximation.

Dimensional Analysis

In engineering and science, dimensional analysis is the analysis of the relationships between different physical quantities by identifying their fundamental dimensions (such as length, mass, time, and electric charge) and units of measure (such as miles vs. kilometers, or pounds vs. kilograms vs. grams) and tracking these dimensions as calculations or comparisons are performed. Converting from one dimensional unit to another is often somewhat complex. Dimensional analysis, or more specifically the factor-label method, also known as the unit-factor method, is a widely used technique for performing such conversions using the rules of algebra.

Any physically meaningful equation (and any inequality and inequation) must have the same dimensions on the left and right sides. Checking this is a common application of performing dimensional analysis. Dimensional analysis is also routinely used as a check on the plausibility of derived equations and computations. It is generally used to categorize types of physical quantities and units based on their relationship to or dependence on other units.

Concrete Numbers and Fundamental Units

Many parameters and measurements in the physical sciences and engineering are expressed as a concrete number - a numerical quantity and a corresponding dimensional unit. Often a quantity is a combination of multiple dimensions; for example, speed is a combination of length and time, e.g. 60 miles per hour or 1.4 km per second. Compound relations with "per" are expressed with division, e.g. 60 mi/1 h. Other relations can involve multiplication (often shown with · or implied), exponents (like m^2 for square meters), or combinations thereof.

Sometimes the names of units obscure their compound nature. For example, an ampere is a measure of electrical current, which is fundamentally electrical charge per unit time and is measured in coulombs (a unit of electrical charge) per second, so 1A = 1C/s. One Newton is kg·m/s^2. A unit of measure which is not compound - it cannot be decomposed into other units - is known as a fundamental unit.

Percentages and Derivatives

Percentages are dimensionless quantities, since they are ratios of two quantities with the same dimensions. In other words, the % sign can be read as "1/100", since 1% = 1/100.

Derivatives with respect to a quantity add the dimensions of the variable one is differentiating with respect to on the denominator. Thus:

- position (x) has units of L (Length);
- derivative of position with respect to time (dx/dt, velocity) has units of L/T – Length from position, Time from the derivative;
- the second derivative (d^2x/dt^2, acceleration) has units of L/T^2.

In economics, one distinguishes between stocks and flows: a stock has units of "units" (say, widgets or dollars), while a flow is a derivative of a stock, and has units of "units/time" (say, dollars/year).

In some contexts, dimensional quantities are expressed as dimensionless quantities or percentages by omitting some dimensions. For example, Debt to GDP ratios are generally expressed as percentages: total debt outstanding (dimension of Currency) divided by annual GDP (dimension of Currency) – but one may argue that in comparing a stock to a flow, annual GDP should have dimensions of Currency/Time (Dollars/Year, for instance), and thus Debt to GDP should have units of years.

Conversion Factor

In dimensional analysis, a ratio which converts one unit of measure into another without changing the quantity is called a conversion factor. For example, kPa and bar are both units of pressure, and 100 kPa = 1 bar. The rules of algebra allow both sides of an equation to be divided by the same expression, so this is equivalent to 100 kPa/1 bar = 1. Since any quantity can be multiplied by 1 without changing it, the expression –100 kPa/1 bar" can be used to convert from bars to kPa by multiplying it with the quantity to be converted, including units. For example, 5 bar * 100 kPa/1 bar = 500 kPa because 5*100/1=500, and bar/bar cancels out, so 5 bar = 500 kPa.

Dimensional Homogeneity

The most basic rule of dimensional analysis is that of dimensional homogeneity. Only *commensurable* quantities (quantities with the same dimensions) may be *compared, equated, added,* or *subtracted.* However, the dimensions form a "multiplicative group" and consequently:

One may take *ratios* of *incommensurable* quantities (quantities with different dimensions), and *multiply* or *divide* them.

For example, it makes no sense to ask if 1 hour is more, the same, or less than 1 kilometer, as these have different dimensions, nor to add 1 hour to 1 kilometer. On the other hand, if an object travels 100 km in 2 hours, one may divide these and conclude that the object's average speed was 50 km/h.

The rule implies that in a physically meaningful *expression* only quantities of the same dimension can be added, subtracted, or compared. For example, if m_{man}, m_{rat} and L_{man} denote, respectively, the mass of some man, the mass of a rat and the length of that man, the dimensionally homogeneous expression $m_{man} + m_{rat}$ is meaningful, but the heterogeneous expression $m_{man} + L_{man}$ is meaningless. However,

m_{man}/L^2_{man} is fine. Thus, dimensional analysis may be used as a sanity check of physical equations: the two sides of any equation must be commensurable or have the same dimensions.

Even when two physical quantities have identical dimensions, it may nevertheless be meaningless to compare or add them. For example, although torque and energy share the dimension ML^2/T^2, they are fundamentally different physical quantities.

To compare, add, or subtract quantities with the same dimensions but expressed in different units, the standard procedure is first to convert them all to the same units. For example, to compare 32 metres with 35 yards, use 1 yard = 0.9144 m to convert 35 yards to 32.004 m.

A related principle is that any physical law that accurately describes the real world must be independent of the units used to measure the physical variables. For example, Newton's laws of motion must hold true whether distance is measured in miles or kilometers. This principle gives rise to the form that conversion factors must take between units that measure the same dimension: multiplication by a simple constant. It also ensures equivalence; for example, if two buildings are the same height in feet, then they must be the same height in meters.

The Factor-Label Method for Converting Units

The factor-label method is the sequential application of conversion factors expressed as fractions and arranged so that any dimensional unit appearing in both the numerator and denominator of any of the fractions can be cancelled out until only the desired set of dimensional units is obtained. For example, 10 miles per hour can be converted to meters per second by using a sequence of conversion factors as shown below:

$$\frac{10\ \cancel{\text{mile}}}{1\ \cancel{\text{hour}}} \times \frac{1609\ \text{meter}}{1\ \cancel{\text{mile}}} \times \frac{1\ \cancel{\text{hour}}}{3600\ \text{second}} = 4.47\ \frac{\text{meter}}{\text{second}}.$$

It can be seen that each conversion factor is equivalent to the value of one. For example, starting with 1 mile = 1609 meters and dividing both sides of the equation by 1 mile yields 1 mile / 1 mile = 1609 meters / 1 mile, which when simplified yields 1 = 1609 meters / 1 mile.

So, when the units *mile* and *hour* are cancelled out and the arithmetic is done, 10 miles per hour converts to 4.47 meters per second.

As a more complex example, the concentration of nitrogen oxides (i.e., NOx) in the flue gas from an industrial furnace can be converted to a mass flow rate expressed in grams per hour (i.e., g/h) of NOx by using the following information as shown below:

NOx concentration

= 10 parts per million by volume = 10 ppmv = 10 volumes/ 10^6 volumes

NOx molar mass

= 46 kg/kgmol (sometimes also expressed as 46 kg/kmol)

Flow rate of flue gas

= 20 cubic meters per minute = 20 m^3/min

The flue gas exits the furnace at 0 °C temperature and 101.325 kPa absolute pressure.

The molar volume of a gas at 0 °C temperature and 101.325 kPa is 22.414 m^3/kgmol.

$$\frac{10\ \cancel{m^3\ NOx}}{10^6\ \cancel{m^3\ gas}} \times \frac{20\ \cancel{m^3\ gas}}{1\ \cancel{minute}} \times \frac{60\ \cancel{minute}}{1\ hour} \times \frac{1\ \cancel{kgmol\ NOx}}{22.414\ \cancel{m^3\ gas}} \times \frac{46\ \cancel{kg\ NOx}}{1 \infty\ \cancel{kgmol\ NOx}} \times \frac{1000\ g}{1\ \cancel{kg}} = 24.63\ \frac{g\ NOx}{hour}$$

After cancelling out any dimensional units that appear both in the numerators and denominators of the fractions in the above equation, the NOx concentration of 10 ppm_v converts to mass flow rate of 24.63 grams per hour.

Checking Equations that Involve Dimensions

The factor-label method can also be used on any mathematical equation to check whether or not the dimensional units on the left hand side of the equation are the same as the dimensional units on the right hand side of the equation. Having the same units on both sides of an equation does not guarantee that the equation is correct, but having different units on the two sides of an equation does guarantee that the equation is wrong.

For example, check the Universal Gas Law equation of $P \cdot V = n \cdot R \cdot T$, when:

- the pressure P is in pascals (Pa)
- the volume V is in cubic meters (m^3)
- the amount of substance n is in moles (mol)
- the universal gas law constant R is 8.3145 Pa $\cdot m^3$/(mol $\cdot$K)
- the temperature T is in kelvins (K)

$$\mathrm{Pa\ m^3} = \frac{\cancel{\mathrm{mol}}}{1} \times \frac{\mathrm{Pa\ m^3}}{\cancel{\mathrm{mol}}\ \cancel{\mathrm{K}}} \times \frac{\cancel{\mathrm{K}}}{1}$$

As can be seen, when the dimensional units appearing in the numerator and denominator of the equation's right hand side are cancelled out, both sides of the equation have the same dimensional units.

Limitations

The factor-label method can convert only unit quantities for which the units are in a linear relationship intersecting at 0. Most units fit this paradigm. An example for which it cannot be used is the conversion between degrees Celsius and kelvins (or Fahrenheit). Between degrees Celsius and kelvins, there is a constant difference rather than a constant ratio, while between Celsius and Fahrenheit there is neither a constant difference nor a constant ratio. There is however an affine transform ($ax+b$), (rather than a linear transform (ax)) between them.

For instance, the freezing point of water is 0° in Celsius and 32° in Fahrenheit, and a 5° change in Celsius corresponds to a 9° change in Fahrenheit. Thus, to convert from Fahrenheit to Celsius one subtracts 32° (displacement from one point), multiplies by 5 and divides by 9 (scales by the ratio of units), and adds 0 (displacement from new point). Reversing this yields the formula for Celsius; one could have started with the equivalence between 100° Celsius and 212° Fahrenheit, though this would yield the same formula at the end.

Hence, to convert Fahrenheit to Celsius, enter the Fahrenheit value into this formula:

°C = (°F-32°) ÷ 1.8

Note that dividing by 1.8 is the same as multiplying by 5 and dividing by 9.

And, to convert Celsius to Fahrenheit, enter the Celsius value into this formula:

°F = 1.8(°C) + 32°

Note that multiplying by 1.8 is the same as multiplying by 9 and dividing by 5.

Applications

Dimensional analysis is most often used in physics and chemistry- and in the mathematics thereof- but finds some applications outside of those fields as well.

Mathematics

A simple application of dimensional analysis to mathematics is in computing the form of the volume of an n-ball (the solid ball in n-dimensions), or the area of its surface, the n-sphere: being an n-dimensional figure, the volume scales as x^n, while the surface area, being $(n-1)$-dimensional, scales as x^{n-1}. Thus the volume of the n-ball in terms of the radius is $C_n r^n$, for some constant C_n. Determining the constant takes more involved mathematics, but the form can be deduced and checked by dimensional analysis alone.

Finance, Economics, and Accounting

In finance, economics, and accounting, dimensional analysis is most commonly referred to in terms of the distinction between stocks and flows. More generally, dimensional analysis is used in interpreting various financial ratios, economics ratios, and accounting ratios.

- For example, the P/E ratio has dimensions of time (units of years), and can be interpreted as "years of earnings to earn the price paid."
- In economics, debt-to-GDP ratio also has units of years (debt has units of currency, GDP has units of currency/year).
- More surprisingly, bond duration also has units of years, which can be shown by dimensional analysis, but takes some financial intuition to understand.
- Velocity of money has units of 1/Years (GDP/Money supply has units of Currency/Year over Currency): how often a unit of currency circulates per year.
- Interest rates are often expressed as a percentage, but more properly percent per annum, which has dimensions of 1/Years.

Critics of mainstream economics, notably including adherents of Austrian economics, have claimed that it lacks dimensional consistency.

Fluid Mechanics

Common dimensionless groups in fluid mechanics include:

- Reynolds number(Re) generally important in all types of fluid problems.

  Re=ρVL/μ

- Froude number(Fr) modelling flow with a free surface.

  Fr=V/sqrt(gL)

- Euler number(Eu) used in problems in which pressure is of interest.

  P/(ρV^2)

Chapter 2

Thermal Energy

Thermal energy is a term sometimes used to refer to the internal energy present in a system in a state of thermodynamic equilibrium by virtue of its temperature. The average translational kinetic energy possessed by free particles in a system of free particles in thermodynamic equilibrium (as measured in the frame of reference of the centree of mass of that system) may also be referred to as the thermal energy per particle.

Microscopically, the thermal energy may include both the kinetic energy and potential energy of a system's constituent particles, which may be atoms, molecules, electrons, or particles. It originates from the individually random, or disordered, motion of particles in a large ensemble. In ideal monatomic gases, thermal energy is entirely kinetic energy. In other substances, in cases where some of thermal energy is stored in atomic vibration or by increased separation of particles having mutual forces of attraction, the thermal energy is equally partitioned between potential energy and kinetic energy. Thermal energy is thus equally partitioned between all available degrees of freedom of the particles. As noted, these degrees of freedom may include pure translational motion in gases, rotational motion, vibrational motion and associated potential energies. In general, due to quantum mechanical reasons, the availability of any such degrees of freedom is a function of the energy in the system, and therefore depends on the temperature.

Macroscopically, the thermal energy of a system at a given temperature is proportional to its heat capacity. However, since the heat capacity differs according to whether or not constant volume or constant pressure is specified, or phase changes permitted, the heat capacity cannot be used to define thermal energy unless it is done in such a way as to insure that only heat gain or loss (not work) makes any changes in the internal energy of the system. Usually, this means specifying the "constant volume heat capacity" of the system so that no work is done. Also the heat capacity of a system for such purposes must not include heat absorbed by any chemical reaction or process.

Differentiation from Heat

Heat, in the strict use in physics, is characteristic only of a process, i.e. it is absorbed or produced as an energy *exchange*, always as a result of a temperature difference. Heat is thermal energy in the process of transfer or conversion across a boundary of one region of matter to another, as a result of a temperature difference. In engineering, the terms "heat" and "heat transfer" are thus used nearly interchangeably, since heat is always understood to be in the process of transfer. The energy transferred by heat is called by other terms (such as thermal energy or latent energy) when this energy is no longer in net transfer, and has become static. Thus, heat is not a static property of matter. Matter does not contain heat, but rather thermal energy, and even the thermal energy is subject to transformations into and out of other types of energy, and so can be considered to be "conserved" only when these processes are small. The heat transfer rate or heating rate is the amount of energy per unit time being transferred as heat, or the heat power.

When two thermodynamic systems with different temperatures are brought into diathermic contact, heat flow occurs from the hotter to the colder system, causing a decrease in the thermal energy of the hotter system, and an increase in the thermal energy of the colder system. Heat flow may cause work to be performed on a system by compressing a system's volume, for example. A heat engine uses the movement of thermal energy (heat flow) to do mechanical work. When two systems have reached a thermodynamic equilibrium, they have attained the same temperature and the net exchange of thermal energy vanishes and heat flow ceases.

Definitions

Thermal energy is the portion of the thermodynamic or internal energy of a system that is responsible for the temperature of the system. The thermal energy of a system scales with its size and is therefore an extensive property. It is not a state function of the system unless the system has been constructed so that all changes in internal energy are due to changes in thermal energy, as a result of heat transfer (not work). Otherwise thermal energy is dependent on the way or method by which the system attained its temperature.

From a macroscopic thermodynamic description, the thermal energy of a system is given by its constant volume specific heat capacity *C(T)*, a temperature coefficient also called thermal capacity, at any given absolute temperature (*T*):

$$U_{thermal} = C(T) \cdot T.$$

The heat capacity is a function of temperature itself, and is typically measured and specified for certain standard conditions and a specific amount of substance (molar heat capacity) or mass units (specific heat capacity). At constant volume (*V*), C_V it is the temperature coefficient of energy. In practice, given a narrow temperature range, for example the operational range of a heat engine, the heat capacity of a system is often constant, and thus thermal energy changes are conveniently measured as temperature fluctuations in the system.

In the microscopical description of statistical physics, the thermal energy is identified with the mechanical kinetic energy of the constituent particles or other forms of kinetic energy associated with quantum-mechanical microstates.

The distinguishing difference between the terms *kinetic energy* and *thermal energy* is that thermal energy is the *mean* energy of disordered, i.e. random, motion of the particles or the oscillations in the system. The conversion of energy of ordered motion to thermal energy results from collisions.

All kinetic energy is partitioned into the degrees of freedom of the system. The average energy of a single particle with *f* quadratic degrees of freedom in a thermal bath of temperature *T* is a statistical mean energy given by the equipartition theorem as

$$E_{thermal} = f \cdot \tfrac{1}{2} kT$$

where *k* is the Boltzmann constant. The total thermal energy of a sample of matter or a thermodynamic system is consequently the average sum of the kinetic energies of all particles in the system. Thus, for a system of *N* particles its thermal energy is

$$U_{thermal} = N \cdot f \cdot \tfrac{1}{2} kT.$$

For gaseous systems, the factor *f*, the number of degrees of freedom, commonly has the value 3 in the case of the monatomic gas, 5 for many diatomic gases, and 7 for larger molecules at ambient temperatures. In general however, it is a function of the temperature of the system as internal modes of motion, vibration, or rotation become available in higher energy regimes.

$U_{thermal}$ is not the total energy of a system. Physical systems also contains static potential energy (such as chemical energy) that arises from interactions between particles, nuclear energy associated with atomic nuclei of particles, and even the rest mass energy due to the equivalence of energy and mass.

Thermal Energy of the Ideal Gas

Thermal energy is most easily defined in the context of the ideal gas, which is well approximated by a monatomic gas at low pressure. The ideal gas is a gas of particles considered as point objects of perfect spherical symmetry that interact only by elastic collisions and fill a volume such that their mean free path between collisions is much larger than their diameter.

The mechanical kinetic energy of a single particle is

$$E_{kinetic} = \tfrac{1}{2}mv^2$$

where m is the particle's mass and v is its velocity. The thermal energy of the gas sample consisting of N atoms is given by the sum of these energies, assuming no losses to the container or the environment:

$$U_{thermal} = \tfrac{1}{2}Nm\overline{v^2} = \tfrac{3}{2}NkT,$$

where the line over the velocity term indicates that the average value is calculated over the entire ensemble. The total thermal energy of the sample is proportional to the macroscopic temperature by a constant factor accounting for the three translational degrees of freedom of each particle and the Boltzmann constant. The Boltzmann constant converts units between the microscopic model and the macroscopic temperature. This formalism is the basic assumption that directly yields the ideal gas law and it shows that for the ideal gas, the internal energy U consists only of its thermal energy:

$$U = U_{thermal}.$$

Historical Context

In an 1847 lecture entitled *On Matter, Living Force, and Heat,* James Prescott Joule characterized various terms that are closely related to thermal energy and heat. He identified the terms latent heat and sensible heat as forms of heat each effecting distinct physical phenomena, namely the potential and kinetic energy of particles, respectively. He describes latent energy as the energy of interaction in a given configuration of particles, i.e. a form of potential energy, and the sensible heat as an energy affecting temperature measured by the thermometer due to the thermal energy, which he called the *living force.*

Distinction of Thermal Energy and Heat

In thermodynamics, heat must always be defined as energy in exchange between two systems, or a single system and its surroundings. According to the zeroth law of thermodynamics, heat is exchanged *between* thermodynamic systems in thermal contact only if their temperatures are different, as this is the condition when the net exchange of thermal energy is non-zero. For the purpose of distinction, a system is defined to be enclosed by a well-characterized boundary. If heat traverses the boundary in direction *into* the system, the internal energy change is considered to be a positive quantity, while *exiting* the system, it is negative. As a process variable, heat is never a property of the system, nor is it *contained* within the boundary of the system.

In contrast to heat, thermal energy exists on both sides of a boundary. It is the statistical mean of the microscopic fluctuations of the kinetic energy of the systems' particles, and it is the source and the effect of the transfer of heat across a system boundary. Statistically, thermal energy is always exchanged between systems, even when the temperatures on both sides is the same, i.e. the systems are in thermal equilibrium. However, at equilibrium, the *net* exchange of thermal energy is zero, and therefore there is no heat.

Thermal energy may be increased in a system by other means than heat, for example when mechanical or electrical work is performed on the system. No qualitative difference exists between the thermal energy added by other means. Thermal energy is not a state function, although it may be closely related to the internal energy of some systems, which is a state function. There is also no need in classical thermodynamics to characterize the thermal energy in terms of atomic or molecular behaviour. A change in thermal energy induced in a system is the product of the change in entropy and the temperature of the system.

Heat exchanged across a boundary may cause changes other than a change in temperature. For example, it may cause phase transitions, such as melting or evaporation, which are changes in the configuration of a material. Since such an energy exchange is not observable by a change in temperature, it is called a latent heat and represents a change in the potential energy of the system.

Rather than being itself the thermal energy involved in a transfer, heat is sometimes also understood as the process of that transfer, i.e. *heat* functions as a verb.

The Origin of Heat Energy on Earth

Earth's proximity to the Sun is the reason that almost everything near Earth's surface is warm with a temperature substantially above absolute zero. Solar radiation constantly replenishes heat energy that Earth loses into space and a relatively stable state of near equilibrium is achieved. Because of the wide variety of heat diffusion mechanisms (one of which is black-body radiation which occurs at the speed of light), objects on Earth rarely vary too far from the global mean surface and air temperature of 287 to 288 K (14 to 15 °C). The more an object's or system's temperature varies from this average, the more rapidly it tends to come back into equilibrium with the ambient environment.

Thermal Energy of Individual Particles

The term *thermal energy* is also often used as a property of single particles to designate the kinetic energy of the particles. An example is the description of thermal neutrons having a certain thermal energy, which means that the kinetic energy of the particle is equivalent to the temperature of its surroundings.

Heat and Temperature

In everyday speech, heat and temperature go hand in hand: the hotter something is, the greater its temperature. However, there is a subtle difference in the way we use the two words in everyday speech, and this subtle difference becomes crucial when studying physics.

Temperature is a property of a material, and thus depends on the material, whereas heat is a form of energy existing on its own. The difference between heat and temperature is analogous to the difference between money and wealth. For example, $200 is an amount of money: regardless of who owns it, $200 is $200. With regard to wealth, though, the significance of $200 varies from person to person. If you are ten and carrying $200 in your wallet, your friends might say you are wealthy or ask to borrow some money. However, if you are thirty-five and carrying $200 in your wallet, your friends will probably not take that as a sign of great wealth, though they may still ask to borrow your money.

Temperature

While temperature is related to thermal energy, there is no absolute correlation between the amount of thermal energy (heat) of an object and its temperature. Temperature measures the concentration of thermal energy in an object in much the same way that density measures the concentration of matter in an object. As a result, a large object will have a much lower temperature than a small object with the same amount of thermal energy. As we shall see shortly, different materials respond to changes in thermal energy with more or less dramatic changes in temperature.

Heat Transfer

Heat transfer describes the exchange of thermal energy, between physical systems depending on the temperature and pressure, by dissipating heat. The fundamental modes of heat transfer are *conduction* or *diffusion, convection* and *radiation.*

The exchange of kinetic energy of particles through the boundary between two systems which are at different temperatures from each other or from their surroundings. Heat transfer always occurs from a region of high temperature to another region of lower temperature. Heat transfer changes the internal energy of both systems involved according to the First Law of Thermodynamics. The Second Law of Thermodynamics defines the concept of thermodynamic entropy, by measurable heat transfer.

Thermal equilibrium is reached when all involved bodies and the surroundings reach the same temperature. Thermal expansion is the tendency of matter to change in volume in response to a change in temperature.

Overview

Heat is defined in physics as the transfer of thermal energy across a well-defined boundary around a thermodynamic system. The thermodynamic free energy is the amount of work that a thermodynamic system can perform. Enthalpy is a thermodynamic potential, designated by the letter "≈, that is the sum of the internal energy of the system (U) plus the product of pressure (P) and volume (V). Joule is a unit to quantify energy, work, or the amount of heat.

Heat transfer is a process function (or path function), as opposed to functions of state; therefore, the amount of heat transferred in a thermodynamic process that changes the state of a system depends on how that process occurs, not only the net difference between the initial and final states of the process.

Thermodynamic and mechanical heat transfer is calculated with the heat transfer coefficient, the proportionality between the heat flux and the thermodynamic driving force for the flow of heat. Heat flux is a quantitative, vectorial representation of the heat flow through a surface.

In engineering contexts, the term *heat* is taken as synonymous to thermal energy. This usage has its origin in the historical interpretation of heat as a fluid (*caloric*) that can be transferred by various causes, and that is also common in the language of laymen and everyday life.

The transport equations for thermal energy (Fourier's law), mechanical momentum (Newton's law for fluids), and mass transfer (Fick's laws of diffusion) are similar, and analogies among these three transport processes have been developed to facilitate prediction of conversion from any one to the others.

Thermal engineering concerns the generation, use, conversion, and exchange of heat transfer. As such, heat transfer is involved in almost every sector of the economy. Heat transfer is classified into various mechanisms, such as thermal conduction, thermal convection, thermal radiation, and transfer of energy by phase changes.

Mechanisms

The fundamental modes of heat transfer are:

Advection: Advection is the transport mechanism of a fluid substance or conserved property from one location to another, depending on motion and momentum.

Conduction or Diffusion: The transfer of energy between objects that are in physical contact. Thermal conductivity is the property of a material to conduct heat and evaluated primarily in terms of Fourier's Law for heat conduction.

Convection: The transfer of energy between an object and its environment, due to fluid motion. The average temperature, is a reference for evaluating properties related to convective heat transfer.

Radiation: The transfer of energy from the movement of charged particles within atoms is converted to electromagnetic radiation.

Advection

By transferring matter, energy—including thermal energy—is moved by the physical transfer of a hot or cold object from one place to another. This can be as simple as placing hot water in a bottle and heating a bed, or the movement of an iceberg in changing ocean currents. A practical example is thermal hydraulics. This can be described by the formula:

$$Q = v \cdot \rho \cdot c_p \cdot \Delta T$$

where Q is heat flux (W/m^2), ρ is density (kg/m^3), c_p is heat capacity at constant pressure (J/(kg*K)), ΔT is the change in temperature (K), v is velocity (m/s).

Conduction

On a microscopic scale, heat conduction occurs as hot, rapidly moving or vibrating atoms and molecules interact with neighbouring atoms and molecules, transferring some of their energy (heat) to these neighbouring particles. In other words, heat is transferred by conduction when adjacent atoms vibrate against one another, or as electrons move from one atom to another. Conduction is the most significant means of heat transfer within a solid or between solid objects in thermal contact. Fluids—especially gases—are less conductive. Thermal contact conductance is the study of heat conduction between solid bodies in contact.

Steady state conduction is a form of conduction that happens when the temperature difference driving the conduction is constant, so that after an equilibration time, the spatial distribution of temperatures in the conducting object does not change any further. In steady state conduction, the amount of heat entering a section is equal to amount of heat coming out.

Transient conduction occurs when the temperature within an object changes as a function of time. Analysis of transient systems is more complex and often calls for the application of approximation theories or numerical analysis by computer.

Convection

The flow of fluid may be forced by external processes, or sometimes (in gravitational fields) by buoyancy forces caused when thermal energy expands the fluid (for example in a fire plume), thus influencing its own transfer. The latter process is often called "natural convection". All convective processes also move heat partly by diffusion, as well. Another form of convection is forced convection. In this case the fluid is forced to flow by use of a pump, fan or other mechanical means.

Convective heat transfer, or convection, is the transfer of heat from one place to another by the movement of fluids, a process that is essentially the transfer of heat via mass transfer. Bulk motion of fluid enhances heat transfer in many physical situations, such as (for example) between a solid surface and the fluid. Convection is usually the dominant form of heat transfer in liquids and gases. Although sometimes discussed as a third method of heat transfer, convection is usually used to describe the combined effects of heat conduction within the fluid (diffusion) and heat transference by bulk fluid flow streaming. The process of transport by fluid streaming is known as advection, but pure advection is a term that is generally associated only with mass transport in fluids, such as advection of pebbles in a river. In the case of heat transfer in fluids, where transport by advection in a fluid is always also accompanied by transport via heat diffusion (also known as heat conduction) the process of heat convection is understood to refer to the sum of heat transport by advection and diffusion/conduction.

Free, or natural, convection occurs when bulk fluid motions (streams and currents) are caused by buoyancy forces that result from density variations due to variations of temperature in the fluid. *Forced* convection is a term used when the streams and currents in the fluid are induced by external means—such as fans, stirrers, and pumps—creating an artificially induced convection current.

Convection-Cooling

Convective cooling is sometimes described as Newton's law of cooling:

The rate of heat loss of a body is proportional to the temperature difference between the body and its surroundings.

However, by definition, the validity of Newton's law of cooling requires that the rate of heat loss from convection be a linear function of ("proportional to") the temperature difference that drives heat transfer, and in convective cooling this is sometimes not the case. In general, convection is not linearly dependent on temperature gradients, and in some cases is strongly nonlinear. In these cases, Newton's law does not apply.

Convection vs. Conduction

In a body of fluid that is heated from underneath its container, conduction and convection can be considered to compete for dominance. If heat conduction is too great, fluid moving down by convection is heated by conduction so fast that its downward movement will be stopped due to its buoyancy, while fluid moving up by convection is cooled by conduction so fast that its driving buoyancy will diminish. On the other hand, if heat conduction is very low, a large temperature gradient may be formed and convection might be very strong.

The Rayleigh number (Ra) is a measure determining the relative strength of conduction and convection.

$$Ra = \frac{g\Delta\rho L^3}{\mu\alpha} = \frac{g\beta\Delta T L^3}{\nu\alpha}$$

where

- g is acceleration due to gravity,

- ρ is the density with $\Delta\rho$ being the density difference between the lower and upper ends,
- μ is the dynamic viscosity,
- α is the Thermal diffusivity,
- β is the volume thermal expansivity (sometimes denoted α elsewhere),
- *T* is the temperature,
- ν is the kinematic viscosity, and
- *L* is characteristic length.

The Rayleigh number can be understood as the ratio between the rate of heat transfer by convection to the rate of heat transfer by conduction; or, equivalently, the ratio between the corresponding timescales (i.e. conduction timescale divided by convection timescale), up to a numerical factor. This can be seen as follows, where all calculations are up to numerical factors depending on the geometry of the system.

The buoyancy force driving the convection is roughly $g\Delta\rho L^3$, so the corresponding pressure is roughly $g\Delta\rho L$. In steady state, this is cancelled by the shear stress due to viscosity, and therefore roughly equals $\mu V / L = \mu / T_{conv}$, where *V* is the typical fluid velocity due to convection and T_{conv} the order of its timescale. The conduction timescale, on the other hand, is of the order of $T_{cond} = L^2 / \alpha$.

Convection occurs when the Rayleigh number is above 1,000–2,000.

Radiation

Thermal radiation occurs through a vacuum or any transparent medium (solid or fluid). It is the transfer of energy by means of photons in electromagnetic waves governed by the same laws. Earth's radiation balance depends on the incoming and the outgoing thermal radiation, Earth's energy budget. Anthropogenic perturbations in the climate system, are responsible for a positive radiative forcing which reduces the net longwave radiation loss out to Space.

Thermal radiation is energy emitted by matter as electromagnetic waves, due to the pool of thermal energy in all matter with a temperature above absolute zero. Thermal radiation propagates without the presence of matter through the vacuum of space.

Thermal radiation is a direct result of the random movements of atoms and molecules in matter. Since these atoms and molecules are composed of charged particles (protons and electrons), their movement results in the emission of electromagnetic radiation, which carries energy away from the surface.

The Stefan-Boltzmann equation, which describes the rate of transfer of radiant energy, is as follows for an object in a vacuum :

$$Q = \epsilon\sigma T^4$$

For radiative transfer between two objects, the equation is as follows:

$$Q = \epsilon\sigma(T_a^4 - T_b^4)$$

where Q is the rate of heat transfer, ε is the emissivity (unity for a black body), ó is the Stefan-Boltzmann constant, and T is the absolute temperature (in Kelvin or Rankine). Radiation is typically only important for very hot objects, or for objects with a large temperature difference.

Radiation from the sun, or solar radiation, can be harvested for heat and power. Unlike conductive and convective forms of heat transfer, thermal radiation can be concentrated in a small spot by using reflecting mirrors, which is exploited in concentrating solar power generation. For example, the sunlight reflected from mirrors heats the PS10 solar power tower and during the day it can heat water to 285 °C (545 °F).

Phase Transition

Phase transition or phase change, takes place in a thermodynamic system from one phase or state of matter to another one by heat transfer. Phase change examples are the melting of ice or the boiling of water. The Mason equation explains the growth of a water droplet based on the effects of heat transport on evaporation and condensation.

Types of phase transition occurring in the four fundamental states of matter, include:

- Solid - Deposition, freezing and solid to solid transformation.
- Gas - Boiling / evaporation, recombination / deionization, and sublimation.
- Liquid - Condensation and melting / fusion.
- Plasma - Ionization.

Boiling

The boiling point of a substance is the temperature at which the vapor pressure of the liquid equals the pressure surrounding the liquid and the liquid evaporates resulting in an abrupt change in vapor volume. Saturation temperature means boiling point. The saturation temperature is the temperature for a corresponding saturation pressure at which a liquid boils into its vapor phase. The liquid can be said to be saturated with thermal energy. Any addition of thermal energy results in a phase transition.

At low temperatures, no boiling occurs and the heat transfer rate is controlled by the usual single-phase mechanisms. As the surface temperature is increased, local boiling occurs and vapor bubbles nucleate, grow into the surrounding cooler fluid, and collapse. This is *sub-cooled nucleate boiling*, and is a very efficient heat transfer mechanism. At high bubble generation rates, the bubbles begin to interfere and the heat flux no longer increases rapidly with surface temperature (this is the departure from nucleate boiling, or DNB).

At high temperatures, the hydrodynamically-quieter regime of film boiling is reached. Heat fluxes across the stable vapor layers are low, but rise slowly with temperature. Any contact between fluid and the surface that may be seen probably leads to the extremely rapid nucleation of a fresh vapor layer ("spontaneous nucleation"). At higher temperatures still, a maximum in the heat flux is reached (the critical heat flux, or CHF).

The Leidenfrost Effect demonstrates how nucleate boiling slows heat transfer due to gas bubbles on the heater's surface. As mentioned, gas-phase thermal conductivity is much lower than liquid-phase thermal conductivity, so the outcome is a kind of "gas thermal barrier".

Condensation

Condensation occurs when a vapor is cooled and changes its phase to a liquid. During condensation, the latent heat of vaporization must be released. The amount of the heat is the same as that absorbed during vaporization at the same fluid pressure.

There are several types of condensation:

- Homogeneous condensation, as during a formation of fog.
- Condensation in direct contact with subcooled liquid.
- Condensation on direct contact with a cooling wall of a heat exchanger: This is the most common mode used in industry:
    - o Filmwise condensation is when a liquid film is formed on the subcooled surface, and usually occurs when the liquid wets the surface.
    - o Dropwise condensation is when liquid drops are formed on the subcooled surface, and usually occurs when the liquid does not wet the surface.

Dropwise condensation is difficult to sustain reliably; therefore, industrial equipment is normally designed to operate in filmwise condensation mode.

Melting

Melting is a physical process that results in the phase transition of a substance from a solid to a liquid. The internal energy of a substance is increased, typically by the application of heat or pressure, resulting in a rise of its temperature to the melting point, at which the ordering of ionic or molecular entities in the solid breaks down to a less ordered state and the solid liquefies. An object that has melted completely is molten. Substances in the molten state generally have reduced viscosity with elevated temperature; an exception to this maxim is the element sulfur, whose viscosity increases to a point due to polymerization and then decreases with higher temperatures in its molten state.

Modelling Approaches

Heat transfer can be modelled in the following ways.

Climate Models

Climate models study the radiant heat transfer by using quantitative methods to simulate the interactions of the atmosphere, oceans, land surface, and ice.

Heat Equation

The heat equation is an important partial differential equation that describes the distribution of heat (or variation in temperature) in a given region over time. In some cases, exact solutions of the equation are available; in other cases the equation must be solved numerically using computational methods.

Lumped System Analysis

Lumped system analysis often reduces the complexity of the equations to one first-order linear differential equation, in which case heating and cooling are described by a simple exponential solution, often referred to as Newton's law of cooling.

System analysis by the lumped capacitance model is a common approximation in transient conduction that may be used whenever heat conduction within an object is much faster than heat conduction across the boundary of the object. This is a method of approximation that reduces one aspect of the transient conduction system—that within the object—to an equivalent steady state system. That is, the method assumes that the temperature within the object is completely uniform, although its value may be changing in time. In this method, the ratio of the conductive heat resistance within the object to the convective heat transfer resistance across the object's boundary, known as the *Biot number*, is calculated. For small Biot numbers, the approximation of *spatially uniform temperature within the object* can be used: it can be presumed that heat transferred into the object has time to uniformly distribute itself, due to the lower resistance to doing so, as compared with the resistance to heat entering the object.

First Law of Thermodynamics

The first law of thermodynamics is a version of the law of conservation of energy, adapted for thermodynamic systems. The law of conservation of energy states that the total energy of an isolated system is constant; energy can be transformed from one form to another, but cannot be created or destroyed. The first law is often formulated by stating that the change in the internal energy of a closed system is equal to the amount of heat supplied to the system, minus the amount of work done by the system on its surroundings. Equivalently, perpetual motion machines of the first kind are impossible.

History

The process of development of the first law of thermodynamics was by way of many tries and mistakes of investigation, over a period of about half a century. The first full statements of the law came in 1850 from Rudolf Clausius and from William Rankine; Rankine's statement was perhaps not quite as clear and distinct as was Clausius'. A main aspect of the struggle was to deal with the previously proposed caloric theory of heat.

Germain Hess in 1840 stated a conservation law for the so-called 'heat of reaction' for chemical reactions. His law was later recognised as a consequence of the first law of thermodynamics, but Hess's statement was not explicitly concerned with the relation between energy exchanges by heat and work.

According to Truesdell (1980), Julius Robert von Mayer in 1841 made a statement that meant that "in a process at constant pressure, the heat used to produce expansion is universally interconvertible with work", but this is not a general statement of the first law.

Original Statements: the "Thermodynamic Approach"

The original nineteenth century statements of the first law of thermodynamics appeared in a conceptual framework in which transfer of energy as heat was taken as a primitive notion, not defined or constructed by the theoretical development of the framework, but rather presupposed as prior to it and already accepted. The primitive notion of heat was taken as empirically established, especially through calorimetry regarded as a subject in its own right, prior to thermodynamics. Jointly primitive with this notion of heat were the notions of empirical temperature and thermal equilibrium. This framework did not presume a concept of energy in general, but regarded it as derived or synthesized from the prior notions of heat and work. By one author, this framework has been called the "thermodynamic" approach.

The first explicit statement of the first law of thermodynamics, by Rudolf Clausius in 1850, referred to cyclic thermodynamic processes.

In all cases in which work is produced by the agency of heat, a quantity of heat is consumed which is proportional to the work done; and conversely, by the expenditure of an equal quantity of work an equal quantity of heat is produced.

Clausius also stated the law in another form, referring to the existence of a function of state of the system, the internal energy, and expressed it in terms of a differential equation for the increments of a thermodynamic process. This equation may described as follows:

In a thermodynamic process involving a closed system, the increment in the internal energy is equal to the difference between the heat accumulated by the system and the work done by it.

Because of its definition in terms of increments, the value of the internal energy of a system is not uniquely defined. It is defined only up to an arbitrary additive constant of integration, which can be adjusted to give arbitrary reference zero levels. This non-uniqueness is in keeping with the abstract mathematical nature of the internal energy. The internal energy is customarily stated relative to a conventionally chosen standard reference state of the system.

The concept of internal energy is considered by Bailyn to be of "enormous interest". Its quantity cannot be immediately measured, but can only be inferred, by differencing actual immediate measurements. Bailyn likens it to the energy states of an atom, that were revealed by Bohr's energy relation $h\nu = E_{n''} - E_{n'}$. In each case, an unmeasurable quantity (the internal energy, the atomic energy level) is revealed by considering the difference of measured quantities (increments of internal energy, quantities of emitted or absorbed radiative energy).

Conceptual Revision: the "Mechanical Approach"

In 1907, George H. Bryan wrote about systems between which there is no transfer of matter (closed systems): "Definition. When energy flows from one system or part of a system to another otherwise than by the performance of mechanical work, the energy so transferred is called *heat*."

Largely through the influence of Max Born, in the twentieth century, this revised conceptual approach to the definition of heat came to be preferred by many writers, including Constantin Carathéodory. It might be called the "mechanical approach". This approach takes as its primitive notion energy transferred as work defined by mechanics. From this, it derives the notions of transfer of energy as heat, and of temperature, as theoretical developments. It regards calorimetry as a derived theory. It has an early origin in the nineteenth century, for example in the work of Helmholtz, but also in the work of many others. For this approach, it is necessary to be sure that if there is transfer of energy associated with transfer of matter, then the transfer of energy other than by transfer of matter is by a physically separate pathway, and independently defined and measured, from the transfer of energy by transfer of matter.

Conceptually Revised Statement, According to the Mechanical Approach

The revised statement of the law takes the notions of adiabatic mechanical work, and of non-adiabatic transfer of energy, as empirically or theoretically established primitive notions. It rests on the primitive notion of *walls*, especially adiabatic walls, presupposed as physically established. Energy can pass such walls as only as adiabatic work, reversibly or irreversibly. If transfer of energy as work is not permitted between them, two systems separated by an adiabatic wall can come to their respective internal mechanical and material thermodynamic equilibrium states completely independently of one another.

The revised statement of the law postulates that a change in the internal energy of a system due to an arbitrary process of interest, that takes the system from its specified initial to its specified final state of internal thermodynamic equilibrium, can be determined through the physical existence of a reference process, for those specified states, that occurs purely through stages of adiabatic work.

The Revised Statement is Then

For a closed system, in any arbitrary process of interest that takes it from an initial to a final state of internal thermodynamic equilibrium, the change of internal energy is the same as that for a reference adiabatic work process that links those two states. This is so regardless of the path of the process of interest, and regardless of whether it is an adiabatic or a non-adiabatic process. The reference adiabatic work process may be chosen arbitrarily from amongst the class of all such processes.

This statement is much less close to the empirical basis than are the original statements, but is often regarded as conceptually parsimonious in that it rests only on the concepts of adiabatic work and of non-adiabatic processes, not on the concepts of transfer of energy as heat and of empirical temperature that are presupposed by the original statements. Largely through the influence of Max Born, it is often regarded as theoretically preferable because of this conceptual parsimony. Born particularly observes that the revised approach avoids thinking in terms of what he calls the "imported engineering" concept of heat engines.

Basing his thinking on the mechanical approach, Born in 1921, and again in 1949, proposed to revise the definition of heat. In particular, he referred to the work of Constantin Carathéodory, who had in 1909 stated the first law without defining quantity of heat. Born's definition was specifically for transfers of energy without transfer of matter, and it has been widely followed in textbooks (examples:). Born observes that a transfer of matter between two systems is accompanied by a transfer of internal energy that cannot be resolved into heat and work components. There can be pathways to other systems, spatially separate from that of the matter transfer, that allow heat and work transfer independent of and simultaneous with the matter transfer. Energy is conserved in such transfers.

Description

The first law of thermodynamics for a closed system was expressed in two ways by Clausius. One way referred to cyclic processes and the inputs and outputs of the system, but did not refer to increments in the internal state of the system. The other way referred to any incremental change in the internal state of the system, and did not expect the process to be cyclic. A cyclic process is one that can be repeated indefinitely often and still eventually leave the system in its original state.

In each repetition of a cyclic process, the work done by the system is proportional to the heat consumed by the system. In a cyclic process in which the system does work on its surroundings, it is necessary that some heat be taken in by the system and some be put out, and the difference is the heat consumed by the system in the process. The constant of proportionality is universal and independent of the system and was measured by James Joule in 1845 and 1847, who described it as the *mechanical equivalent of heat.*

For a closed system, in any process, the change in the internal energy is considered due to a combination of heat added to the system and work done *by* the system. Taking ΔU as a change in internal energy, one writes

$$\Delta U = Q - W \text{ (sign convention of Clausius and generally in this article),}$$

where Q and W are quantities of heat supplied to the system by its surroundings and of work done by the system on its surroundings, respectively. This sign convention is implicit in Clausius' statement of the law given above, and is consistent with the use of thermodynamics to study heat engines, which provide useful work that is regarded as positive.

In modern style of teaching science, however, it is conventional to use the IUPAC convention by which the first law is formulated in terms of the work done on the system. With this alternate sign convention for work, the first law for a closed system may be written:

$$\Delta U = Q + W \text{ (sign convention of IUPAC).}$$

This convention follows physicists such as Max Planck, and considers all net energy transfers to the system as positive and all net energy transfers from the system as negative, irrespective of any use for the system as an engine or other device.

When a system expands in a fictive quasistatic process, the work done by the system on the environment is the product, $P\,\mathrm{d}V$, of pressure, P, and volume change, $\mathrm{d}V$, whereas the work done *on* the system is $-P\,\mathrm{d}V$. Using either sign convention for work, the change in internal energy of the system is:

$$\mathrm{d}U = \delta Q - P\mathrm{d}V \text{ (quasi-static process),}$$

where δQ denotes the infinitesimal increment of heat supplied to the system from its surroundings.

Work and heat are expressions of actual physical processes of supply or removal of energy, while the internal energy U is a mathematical abstraction that keeps account of the exchanges of energy that befall the system. Thus the term heat for Q means "that amount of energy added or removed by conduction of heat or by thermal radiation", rather than referring to a form of energy within the system. Likewise, the term work energy for W means "that amount of energy gained or lost as the result of work". Internal energy is a property of the system whereas work done and heat supplied are not. A significant result of this distinction is that a given internal energy change ΔU can be achieved by, in principle, many combinations of heat and work.

Various Statements of the Law for Closed Systems

The law is of very great importance and generality and is consequently thought of from several points of view. Most careful textbook statements of the law express it for closed systems. It is stated in several ways, sometimes even by the same author.

For the thermodynamics of closed systems, the distinction between transfers of energy as work and as heat is central and is within the scope of the present article. For the thermodynamics of open systems, such a distinction is beyond the scope of the present article, but some limited comments are made on it in the section below headed 'First law of thermodynamics for open systems'.

There are two main ways of stating a law of thermodynamics, physically or mathematically. They should be logically coherent and consistent with one another.

An example of a physical statement is that of Planck (1897/1903):

It is in no way possible, either by mechanical, thermal, chemical, or other devices, to obtain perpetual motion, i.e. it is impossible to construct an engine which will work in a cycle and produce continuous work, or kinetic energy, from nothing.

This physical statement is restricted neither to closed systems nor to systems with states that are strictly defined only for thermodynamic equilibrium; it has meaning also for open systems and for systems with states that are not in thermodynamic equilibrium.

An example of a mathematical statement is that of Crawford (1963):

For a given system we let ΔE^{kin} = large-scale mechanical energy, ΔE^{pot} = large-scale potential energy, and ΔE^{tot} = total energy. The first two quantities are specifiable in terms of appropriate mechanical variables, and by definition

$$E^{\text{tot}} = E^{\text{kin}} + E^{\text{pot}} + U .$$

For any finite process, whether reversible or irreversible,

$$\Delta E^{\text{tot}} = \Delta E^{\text{kin}} + \Delta E^{\text{pot}} + \Delta U .$$

The first law in a form that involves the principle of conservation of energy more generally is

$$\Delta E^{\text{tot}} = Q + W .$$

Here Q and W are heat and work added, with no restrictions as to whether the process is reversible, quasistatic, or irreversible.[Warner, *Am. J. Phys.*, 29, 124 (1961)]

This statement by Crawford, for *W*, uses the sign convention of IUPAC, not that of Clausius. Though it does not explicitly say so, this statement refers to closed systems, and to internal energy *U* defined for bodies in states of thermodynamic equilibrium, which possess well-defined temperatures.

The history of statements of the law for closed systems has two main periods, before and after the work of Bryan (1907), of Carathéodory (1909), and the approval of Carathéodory's work given by Born (1921). The earlier traditional versions of the law for closed systems are nowadays often considered to be out of date.

Carathéodory's celebrated presentation of equilibrium thermodynamics refers to closed systems, which are allowed to contain several phases connected by internal walls of various kinds of impermeability and permeability (explicitly including walls that are permeable only to heat). Carathéodory's 1909 version of the first law of thermodynamics was stated in an axiom which refrained from defining or mentioning temperature or quantity of heat transferred. That axiom stated that the internal energy of a phase in equilibrium is a function of state, that the sum of the internal energies of the phases is the total internal energy of the system, and that the value of the total internal energy of the system is changed by the amount of work done adiabatically on it, considering work as a form of energy. That article considered this statement to be an expression of the law of conservation of energy for such systems. This version is nowadays widely accepted as authoritative, but is stated in slightly varied ways by different authors.

The 1909 Carathéodory statement of the law in axiomatic form does not mention heat or temperature, but the equilibrium states to which it refers are explicitly defined by variable sets that necessarily include "non-deformation variables", such as pressures, which, within reasonable restrictions, can be rightly interpreted as empirical temperatures, and the walls connecting the phases of the system are explicitly defined as possibly impermeable to heat or permeable only to heat.

According to Münster (1970), "A somewhat unsatisfactory aspect of Carathéodory's theory is that a consequence of the Second Law must be considered at this point [in the statement of the first law], i.e. that it is not always possible to reach any state 2 from any other state 1 by means of an adiabatic process." Münster instances that no adiabatic process can reduce the internal energy of a system at constant volume. Carathéodory's paper asserts that its statement of the first law corresponds exactly to Joule's experimental arrangement, regarded as an instance of adiabatic work. It does not point out that Joule's experimental arrangement performed essentially irreversible work, through friction of paddles in a liquid, or passage of electric current through a resistance inside the system, driven by motion of a coil and inductive heating, or by an external current source, which can access the system only by the passage of electrons, and so is not strictly adiabatic, because electrons are a form of matter, which cannot penetrate adiabatic walls. The paper goes on to base its main argument on the possibility of quasi-static adiabatic work, which is essentially reversible. The paper asserts that it will avoid reference to Carnot cycles, and then proceeds to base its argument on cycles of forward and backward quasi-static adiabatic stages, with isothermal stages of zero magnitude.

Some respected modern statements of the first law for closed systems assert the existence of internal energy as a function of state defined in terms of adiabatic work and accept the idea that heat is not defined in its own right, that is to say calorimetrically or as due to temperature difference; they define heat as a residual difference between change of internal energy and work done on the system, when that work does not account for the whole of the change of internal energy and the system is not adiabatically isolated.

Sometimes the concept of internal energy is not made explicit in the statement.

Sometimes the existence of the internal energy is made explicit but work is not explicitly mentioned in the statement of the first postulate of thermodynamics. Heat supplied is then defined as the residual change in internal energy after work has been taken into account, in a non-adiabatic process.

A respected modern author states the first law of thermodynamics as "Heat is a form of energy", which explicitly mentions neither internal energy nor adiabatic work. Heat is defined as energy transferred by thermal contact with a reservoir, which has a temperature, and is generally so large that addition and removal of heat do not alter its temperature. A current student text on chemistry defines heat thus: "*heat* is the exchange of thermal energy between a system and its surroundings caused by a temperature difference." The author then explains how heat is defined or measured by calorimetry, in terms of heat capacity, specific heat capacity, molar heat capacity, and temperature.

A respected text disregards the Carathéodory's exclusion of mention of heat from the statement of the first law for closed systems, and admits heat calorimetrically defined along with work and internal energy. Another respected text defines heat exchange as determined by temperature difference, but also mentions that the Born (1921) version is "completely rigorous". These versions follow the traditional approach that is now considered out of date, exemplified by that of Planck (1897/1903).

Evidence for the First Law of Thermodynamics for Closed Systems

The first law of thermodynamics for closed systems was originally induced from empirically observed evidence, including calorimetric evidence. It is nowadays, however, taken to provide the definition of heat via the law of conservation of energy and the definition of work in terms of changes in the external parameters of a system. The original discovery of the law was gradual over a period of perhaps half a century or more, and some early studies were in terms of cyclic processes.

The following is an account in terms of changes of state of a closed system through compound processes that are not necessarily cyclic. This account first considers processes for which the first law is easily verified because of their simplicity, namely adiabatic processes (in which there is no transfer as heat) and adynamic processes (in which there is no transfer as work).

Adiabatic Processes

In an adiabatic process, there is transfer of energy as work but not as heat. For all adiabatic process that takes a system from a given initial state to a given final state, irrespective of how the work is done, the respective eventual total quantities of energy transferred as work are one and the same, determined just by the given initial and final states. The work done on the system is defined and measured by changes in mechanical or quasi-mechanical variables external to the system. Physically, adiabatic transfer of energy as work requires the existence of adiabatic enclosures.

For instance, in Joule's experiment, the initial system is a tank of water with a paddle wheel inside. If we isolate the tank thermally, and move the paddle wheel with a pulley and a weight, we can relate the increase in temperature with the distance descended by the mass. Next, the system is returned to its initial state, isolated again, and the same amount of work is done on the tank using different devices (an electric motor, a chemical battery, a spring,...). In every case, the amount of work can be measured independently. The return to the initial state is not conducted by doing adiabatic work on the system. The evidence shows that the final state of the water (in particular, its temperature and volume) is the same in every case. It is irrelevant if the work is electrical, mechanical, chemical,... or if done suddenly or slowly, as long as it is performed in an adiabatic way, that is to say, without heat transfer into or out of the system.

Evidence of this kind shows that to increase the temperature of the water in the tank, the qualitative kind of adiabatically performed work does not matter. No qualitative kind of adiabatic work has ever been observed to decrease the temperature of the water in the tank.

A change from one state to another, for example an increase of both temperature and volume, may be conducted in several stages, for example by externally supplied electrical work on a resistor in the body, and adiabatic expansion allowing the body to do work on the surroundings. It needs to be shown that the time order of the stages, and their relative magnitudes, does not affect the amount of adiabatic work that needs to be done for the change of state. According to one respected scholar: "Unfortunately, it does not seem that experiments of this kind have ever been carried out carefully. ... We must therefore admit that the statement which we have enunciated here, and which is equivalent to the first law of thermodynamics, is not well founded on direct experimental evidence."

This kind of evidence, of independence of sequence of stages, combined with the above-mentioned evidence, of independence of qualitative kind of work, would show the existence of a very important state variable that corresponds with adiabatic work, but not that such a state variable represented a conserved quantity. For the latter, another step of evidence is needed, which may be related to the concept of reversibility, as mentioned below.

That very important state variable was first recognised and denoted *U* by Clausius in 1850, but he did not then name it, and he defined it in terms not only of work but also of heat transfer in the same process. It was also independently recognised in 1850 by Rankine, who also denoted it *U*; and in 1851 by Kelvin who then called it "mechanical energy", and later "intrinsic energy". In 1865, after some hestitation, Clausius began calling his state function *U* "energy". In 1882 it was named as the *internal energy* by Helmholtz. If only adiabatic processes were of interest, and heat could be ignored, the concept of internal energy would hardly arise or be needed. The relevant physics would be largely covered by the concept of potential energy, as was intended in the 1847 paper of Helmholtz on the principle of conservation of energy, though that did not deal with forces that cannot be described by a potential, and thus did not fully justify the principle. Moreover, that paper was very critical of the early work of Joule that had by then been performed. A great merit of the internal energy concept is that it frees thermodynamics from a restriction to cyclic processes, and allows a treatment in terms of thermodynamic states.

In an adiabatic process, adiabatic work takes the system either from a reference state O with internal energy $U(O)$ to an arbitrary one A with internal energy $U(A)$, or from the state A to the state O:

$$U(A) = U(O) - W_{O \to A}^{\text{adiabatic}} \text{ or } U(O) = U(A) - W_{A \to O}^{\text{adiabatic}}.$$

Except under the special, and strictly speaking, fictional, condition of reversibility, only one of the processes adiabatic, $O \to A$ or adiabatic, $A \to O$ is empirically feasible by a simple application of externally supplied work. The reason for this is given as the second law of thermodynamics and is not considered in the present article.

The fact of such irreversibility may be dealt with in two main ways, according to different points of view:

- Since the work of Bryan (1907), the most accepted way to deal with it nowadays, followed by Carathéodory, is to rely on the previously established concept of quasi-static processes, as follows. Actual physical processes of transfer of energy as work are always at least to some degree irreversible. The irreversibility is often due to mechanisms known as dissipative, that transform bulk kinetic energy into internal energy. Examples are friction and viscosity. If the process is performed more slowly, the frictional or viscous dissipation is less. In the limit of infinitely slow performance, the dissipation tends to zero and then the limiting process, though fictional rather than actual, is notionally reversible, and is called quasi-static. Throughout the course of the fictional limiting quasi-static process, the internal intensive variables of the system are equal to the external intensive variables, those that describe the reactive forces exerted by the surroundings. This can be taken to justify the formula

  (1) $W_{A \to O}^{\text{adiabatic, quasi-static}} = -W_{O \to A}^{\text{adiabatic, quasi-static}}.$

- Another way to deal with it is to allow that experiments with processes of heat transfer to or from the system may be used to justify the formula (1) above. Moreover, it deals to some extent with the problem of lack of direct experimental evidence that the time order of stages of a process does not matter in the determination of internal energy. This way does not provide theoretical purity in terms of adiabatic work processes, but is empirically feasible, and is in accord with experiments actually done, such as the Joule experiments mentioned just above, and with older traditions.

The formula (1) above allows that to go by processes of quasi-static adiabatic work from the state A to the state B we can take a path that goes through the reference state O, since the quasi-static adiabatic work is independent of the path

$$-W_{A\to B}^{\text{adiabatic,quasi-static}} = -W_{A\to O}^{\text{adiabatic,quasi-static}} - W_{O\to B}^{\text{adiabatic,quasi-static}} = W_{O\to A}^{\text{adiabatic,quasi-static}}$$

$$-W_{O\to B}^{\text{adiabatic,quasi-static}} = -U(A) + U(B) = \Delta U$$

This kind of empirical evidence, coupled with theory of this kind, largely justifies the following statement:

For all adiabatic processes between two specified states of a closed system of any nature, the net work done is the same regardless the details of the process, and determines a state function called internal energy, U."

Adynamic Processes

A complementary observable aspect of the first law is about heat transfer. Adynamic transfer of energy as heat can be measured empirically by changes in the surroundings of the system of interest by calorimetry. This again requires the existence of adiabatic enclosure of the entire process, system and surroundings, though the separating wall between the surroundings and the system is thermally conductive or radiatively permeable, not adiabatic. A calorimeter can rely on measurement of sensible heat, which requires the existence of thermometers and measurement of temperature change in bodies of known sensible heat capacity under specified conditions; or it can rely on the measurement of latent heat, through measurement of masses of material that change phase, at temperatures fixed by the occurrence of phase changes under specified conditions in bodies of known latent heat of phase change. The calorimeter can be calibrated by adiabatically doing externally determined work on it. The most accurate method is by passing an electric current from outside through a resistance inside the calorimeter. The calibration allows comparison of calorimetric measurement of quantity of heat transferred with quantity of energy transferred as work. According to one textbook, "The most common device for measuring ΔU is an adiabatic bomb calorimeter." According to another textbook, "Calorimetry is widely used in present day laboratories." According to one opinion, "Most thermodynamic data come from calorimetry..." According to another opinion, "The most common method of measuring "heat" is with a calorimeter."

When the system evolves with transfer of energy as heat, without energy being transferred as work, in an adynamic process, the heat transferred to the system is equal to the increase in its internal energy:

$$Q_{A\to B}^{\text{adynamic}} = \Delta U.$$

General Case for Reversible Processes

Heat transfer is practically reversible when it is driven by practically negligibly small temperature gradients. Work transfer is practically reversible when it occurs so slowly that there are no frictional effects within the system; frictional effects outside the system should also be zero if the process is to be globally reversible. For a particular reversible process in general, the work done reversibly on the system, $W_{A\to B}^{\text{path}\,P_0\text{,reversible}}$, and the heat transferred reversibly to the system, $Q_{A\to B}^{\text{path}\,P_0\text{,reversible}}$ are not required to occur respectively adiabatically or adynamically, but they must belong to the same particular process defined by its particular reversible path, P_0, through the space of thermodynamic states. Then the work and heat transfers can occur and be calculated simultaneously.

Putting the two complementary aspects together, the first law for a particular reversible process can be written

$$-W_{A\to B}^{\text{path}\,P_0\text{,reversible}} + Q_{A\to B}^{\text{path}\,P_0\text{,reversible}} = \Delta U.$$

This combined statement is the expression the first law of thermodynamics for reversible processes for closed systems.

In particular, if no work is done on a thermally isolated closed system we have

$$\Delta U = 0\,.$$

This is one aspect of the law of conservation of energy and can be stated:

The internal energy of an isolated system remains constant.

General Case for Irreversible Processes

If, in a process of change of state of a closed system, the energy transfer is not under a practically zero temperature gradient and practically frictionless, then the process is irreversible. Then the heat and work transfers may be difficult to calculate, and irreversible thermodynamics is called for. Nevertheless, the first law still holds and provides a check on the measurements and calculations of the work done irreversibly on the system, $W_{A \to B}^{\text{path } P_1, \text{irreversible}}$, and the heat transferred irreversibly to the system, $Q_{A \to B}^{\text{path } P_1, \text{irreversible}}$, which belong to the same particular process defined by its particular irreversible path, P_1, through the space of thermodynamic states.

$$-W_{A \to B}^{\text{path } P_1, \text{irreversible}} + Q_{A \to B}^{\text{path } P_1, \text{irreversible}} = \Delta U.$$

This means that the internal energy U is a function of state and that the internal energy change ΔU between two states is a function only of the two states.

Overview of the Weight of Evidence for the Law

The first law of thermodynamics is very general and makes so many predictions that they can hardly all be directly tested by experiment. Nevertheless, very very many of its predictions have been found empirically accurate. And very importantly, no accurately and properly conducted experiment has ever detected a violation of the law. Consequently, within its scope of applicability, the law is so reliably established, that, nowadays, rather than experiment being considered as testing the accuracy of the law, it is far more practical and realistic to think of the law as testing the accuracy of experiment. An experimental result that seems to violate the law may be assumed to be inaccurate or wrongly conceived, for example due to failure to consider an important physical factor.

State Functional Formulation for Infinitesimal Processes

When the heat and work transfers in the equations above are infinitesimal in magnitude, they are often denoted by δ, rather than exact differentials denoted by "d", as a reminder that heat and work do not describe the *state* of any system. The integral of an inexact differential depends upon the particular path taken through the space of thermodynamic parameters while the integral of an exact differential depends only upon the initial and final states. If the initial and final states are the same, then the integral of an inexact differential may or may not be zero, but the integral of an exact differential is always zero. The path taken by a thermodynamic system through a chemical or physical change is known as a thermodynamic process.

For a homogeneous system, with a well-defined temperature and pressure, the expression for dU can be written in terms of exact differentials, if the work that the system does is equal to its pressure times the infinitesimal increase in its volume. Here one assumes that the changes are quasistatic, so slow that there is at each instant negligible departure from thermodynamic equilibrium within the system. In other words, $\delta W = -PdV$ where P is pressure and V is volume. As such a quasistatic process in a homogeneous system is reversible, the total amount of heat added to a closed system can be expressed as $\delta Q = TdS$ where T is the temperature and S the entropy of the system. Therefore, for closed, homogeneous systems:

$$dU = TdS - PdV.$$

The above equation is known as the fundamental thermodynamic relation, for which the independent variables are taken as S and V, with respect to which T and P are partial derivatives of U. While this has been derived for quasistatic changes, it is valid in general, as U can be considered as a thermodynamic state function of the independent variables S and V.

As an example, one may suppose that the system is initially in a state of thermodynamic equilibrium defined by S and V. Then the system is suddenly perturbed so that thermodynamic equilibrium breaks down and no temperature and pressure can be defined. Eventually the system settles down again to a state of thermodynamic equilibrium, defined by an entropy and a volume that differ infinitesimally from the initial values. The infinitesimal difference in internal energy between the initial and final state satisfies the above equation. But the work done and heat added to the system do not satisfy the above expressions. Rather, they satisfy the inequalities: $\delta Q < TdS'$ *and* $\delta W < PdV'$.

In the case of a closed system in which the particles of the system are of different types and, because chemical reactions may occur, their respective numbers are not necessarily constant, the expression for dU becomes:

$$dU = \delta Q - \delta W + \sum_i \mu_i dN_i$$

where dN_i is the (small) increase in amount of type-i particles in the reaction, and μ_i is known as the chemical potential of the type-i particles in the system. If dN_i is expressed in mol then μ_i is expressed in J/mol. The statement of the first law, using exact differentials is now:

$$dU = TdS - PdV + \sum_i \mu_i dN_i.$$

If the system has more external mechanical variables than just the volume that can change, the fundamental thermodynamic relation generalizes to:

$$dU = TdS - \sum_i X_i dx_i + \sum_j \mu_j dN_j.$$

Here the X_i are the generalized forces corresponding to the external variables x_i. The parameters X_i are independent of the size of the system and are called intensive parameters and the x_i are proportional to the size and called extensive parameters.

For an open system, there can be transfers of particles as well as energy into or out of the system during a process. For this case, the first law of thermodynamics still holds, in the form that the internal energy is a function of state and the change of internal energy in a process is a function only of its initial and final states, as noted in the section below headed First law of thermodynamics for open systems.

A useful idea from mechanics is that the energy gained by a particle is equal to the force applied to the particle multiplied by the displacement of the particle while that force is applied. Now consider the first law without the heating term: $dU = -PdV$. The pressure P can be viewed as a force (and in fact has units of force per unit area) while dV is the displacement (with units of distance times area). We may say, with respect to this work term, that a pressure difference forces a transfer of volume, and that the product of the two (work) is the amount of energy transferred out of the system as a result of the process. If one were to make this term negative then this would be the work done on the system.

It is useful to view the TdS term in the same light: here the temperature is known as a "generalized" force (rather than an actual mechanical force) and the entropy is a generalized displacement.

Similarly, a difference in chemical potential between groups of particles in the system drives a chemical reaction that changes the numbers of particles, and the corresponding product is the amount of chemical potential energy transformed in process. For example, consider a system consisting of two phases: liquid water and water vapor. There is a generalized "force" of evaporation that drives water molecules out of the liquid. There is a generalized "force" of condensation that drives vapor molecules out of the vapor. Only when these two "forces" (or chemical potentials) are equal is there equilibrium, and the net rate of transfer zero.

The two thermodynamic parameters that form a generalized force-displacement pair are called "conjugate variables". The two most familiar pairs are, of course, pressure-volume, and temperature-entropy.

Spatially Inhomogeneous Systems

Classical thermodynamics is initially focused on closed homogeneous systems (e.g. Planck 1897/1903), which might be regarded as 'zero-dimensional' in the sense that they have no spatial variation. But it is desired to study also systems with distinct internal motion and spatial inhomogeneity. For such systems, the principle of conservation of energy is expressed in terms not only of internal energy as defined for homogeneous systems, but also in terms of kinetic energy and potential energies of parts of the inhomogeneous system with respect to each other and with respect to long-range external forces. How the total energy of a system is allocated between these three more specific kinds of energy varies according to the purposes of different writers; this is because these components of energy are to some extent mathematical artefacts rather than actually measured physical quantities. For any closed homogeneous component of an inhomogeneous closed system, if E denotes the total energy of that component system, one may write

$$E = E^{\mathrm{kin}} + E^{\mathrm{pot}} + U$$

where E^{kin} and E^{pot} denote respectively the total kinetic energy and the total potential energy of the component closed homogeneous system, and U denotes its internal energy.

Potential energy can be exchanged with the surroundings of the system when the surroundings impose a force field, such as gravitational or electromagnetic, on the system.

A compound system consisting of two interacting closed homogeneous component subsystems has a potential energy of interaction E_{12}^{pot} between the subsystems. Thus, in an obvious notation, one may write

$$E = E_1^{\mathrm{kin}} + E_1^{\mathrm{pot}} + U_1 + E_2^{\mathrm{kin}} + E_2^{\mathrm{pot}} + U_2 + E_{12}^{\mathrm{pot}}$$

The quantity E_{12}^{pot} in general lacks an assignment to either subsystem in a way that is not arbitrary, and this stands in the way of a general non-arbitrary definition of transfer of energy as work. On occasions, authors make their various respective arbitrary assignments.

The distinction between internal and kinetic energy is hard to make in the presence of turbulent motion within the system, as friction gradually dissipates macroscopic kinetic energy of localised bulk flow into molecular random motion of molecules that is classified as internal energy. The rate of dissipation by friction of kinetic energy of localised bulk flow into internal energy, whether in turbulent or in streamlined flow, is an important quantity in non-equilibrium thermodynamics. This is a serious difficulty for attempts to define entropy for time-varying spatially inhomogeneous systems.

First Law of Thermodynamics for Open Systems

For the first law of thermodynamics, there is no trivial passage of physical conception from the closed system view to an open system view. For closed systems, the concepts of an adiabatic enclosure and of an adiabatic wall are fundamental. Matter and internal energy cannot permeate or penetrate such a wall. For an open system, there is a wall that allows penetration by matter. In general, matter in diffusive motion carries with it some internal energy, and some microscopic potential energy changes accompany the motion. An open system is not adiabatically enclosed.

There are some cases in which a process for an open system can, for particular purposes, be considered as if it were for a closed system. In an open system, by definition hypothetically or potentially, matter can pass between the system and its surroundings. But when, in a particular case, the process of interest involves only hypothetical or potential but no actual passage of matter, the process can be considered as if it were for a closed system.

Internal Energy for an Open System

Since the revised and more rigorous definition of the internal energy of a closed system rests upon the possibility of processes by which adiabatic work takes the system from one state to another, this leaves a problem for the definition of internal energy for an open system, for which adiabatic work is not in general possible. According to Max Born, the transfer of matter and energy across an open connection "cannot be reduced to mechanics". In contrast to the case of closed systems, for open systems, in the presence of diffusion, there is no unconstrained and unconditional physical distinction between convective transfer of internal energy by bulk flow of matter, the transfer of internal energy without transfer of matter (usually called heat conduction and work transfer), and change of various potential energies. The older traditional way and the conceptually revised (Carathéodory) way agree that there is no physically unique definition of heat and work transfer processes between open systems.

In particular, between two otherwise isolated open systems an adiabatic wall is by definition impossible. This problem is solved by recourse to the principle of conservation of energy. This principle allows a composite isolated system to be derived from two other component non-interacting isolated systems, in such a way that the total energy of the composite isolated system is equal to the sum of the total energies of the two component isolated systems. Two previously isolated systems can be subjected to the thermodynamic operation of placement between them of a wall permeable to matter and energy, followed by a time for establishment of a new thermodynamic state of internal equilibrium in the new single unpartitioned system. The internal energies of the initial two systems and of the final new system, considered respectively as closed systems as above, can be measured. Then the law of conservation of energy requires that

$$\Delta U_s + \Delta U_o = 0,$$

where ΔU_s and ΔU_o denote the changes in internal energy of the system and of its surroundings respectively. This is a statement of the first law of thermodynamics for a transfer between two otherwise isolated open systems, that fits well with the conceptually revised and rigorous statement of the law stated above.

For the thermodynamic operation of adding two systems with internal energies U_1 and U_2, to produce a new system with internal energy U, one may write $U = U_1 + U_2$; the reference states for U, U_1 and U_2 should be specified accordingly, maintaining also that the internal energy of a system be proportional to its mass, so that the internal energies are extensive variables.

There is a sense in which this kind of additivity expresses a fundamental postulate that goes beyond the simplest ideas of classical closed system thermodynamics; the extensivity of some variables is not obvious, and needs explicit expression; indeed one author goes so far as to say that it could be recognised as a fourth law of thermodynamics, though this is not repeated by other authors.

Also of Course

$$\Delta N_s + \Delta N_o = 0,$$

where ΔN_s and ΔN_o denote the changes in mole number of a component substance of the system and of its surroundings respectively. This is a statement of the law of conservation of mass.

Process of Transfer of Matter between an Open System and its Surroundings

A system connected to its surroundings only through contact by a single permeable wall, but otherwise isolated, is an open system. If it is initially in a state of contact equilibrium with a surrounding subsystem, a thermodynamic process of transfer of matter can be made to occur between them if the surrounding subsystem is subjected to some thermodynamic operation, for example, removal of a partition between it and some further surrounding subsystem. The removal of the partition in the surroundings initiates a process of exchange between the system and its contiguous surrounding subsystem.

An example is evaporation. One may consider an open system consisting of a collection of liquid, enclosed except where it is allowed to evaporate into or to receive condensate from its vapor above it, which may be considered as its contiguous surrounding subsystem, and subject to control of its volume and temperature.

A thermodynamic process might be initiated by a thermodynamic operation in the surroundings, that mechanically increases in the controlled volume of the vapor. Some mechanical work will be done within the surroundings by the vapor, but also some of the parent liquid will evaporate and enter the vapor collection which is the contiguous surrounding subsystem. Some internal energy will accompany the vapor that leaves the system, but it will not make sense to try to uniquely identify part of that internal energy as heat and part of it as work. Consequently, the energy transfer that accompanies the transfer of matter between the system and its surrounding subsystem cannot be uniquely split into heat and work transfers to or from the open system. The component of total energy transfer that accompanies the transfer of vapor into the surrounding subsystem is customarily called 'latent heat of evaporation', but this use of the word heat is a quirk of customary historical language, not in strict compliance with the thermodynamic definition of transfer of energy as heat. In this example, kinetic energy of bulk flow and potential energy with respect to long-range external forces such as gravity are both considered to be zero. The first law of thermodynamics refers to the change of internal energy of the open system, between its initial and final states of internal equilibrium.

Open System with Multiple Contacts

An open system can be in contact equilibrium with several other systems at once.

This includes cases in which there is contact equilibrium between the system, and several subsystems in its surroundings, including separate connections with subsystems through walls that are permeable to the transfer of matter and internal energy as heat and allowing friction of passage of the transferred matter, but immovable, and separate connections through adiabatic walls with others, and separate connections through diathermic walls impermeable to matter with yet others. Because there are physically separate connections that are permeable to energy but impermeable to matter, between the system and its surroundings, energy transfers between them can occur with definite heat and work characters. Conceptually essential here is that the internal energy transferred with the transfer of matter is measured by a variable that is mathematically independent of the variables that measure heat and work.

With such independence of variables, the total increase of internal energy in the process is then determined as the sum of the internal energy transferred from the surroundings with the transfer of matter through the walls that are permeable to it, and of the internal energy transferred to the system as heat through the diathermic walls, and of the energy transferred to the system as work through the adiabatic walls, including the energy transferred to the system by long-range forces. These simultaneously transferred quantities of energy are defined by events in the surroundings of the system. Because the internal energy transferred with matter is not in general uniquely resolvable into heat and work components, the total energy transfer cannot in general be uniquely resolved into heat and work components. Under these conditions, the following formula can describe the process in terms of externally defined thermodynamic variables, as a statement of the first law of thermodynamics:

(2) $\Delta U_0 = Q - W - \sum_{i=1}^{m} \Delta U_i$ (suitably defined surrounding subsystems, general process, quasi-static or irreversible),

where ΔU_0 denotes the change of internal energy of the system, and ΔU_i denotes the change of internal energy of the ith of the m surrounding subsystems that are in open contact with the system, due to transfer between the system and that ith surrounding subsystem, and Q denotes the internal energy transferred as heat from the heat reservoir of the surroundings to the system, and W denotes the energy transferred from the system to the surrounding subsystems that are in adiabatic connection with it. The case of a wall that is permeable to matter and can move so as to allow transfer of energy as work is not considered here.

Combination of First and Second Laws

If the system is described by the energetic fundamental equation, $U_0 = U_0(S, V, N_j)$, and if the process can be described in the quasi-static formalism, in terms of the internal state variables of the system, then the process can also be described by a combination of the first and second laws of thermodynamics, by the formula

$$(3)\ \mathrm{d}U_0 = T\mathrm{d}S - P\mathrm{d}V + \sum_{j=1}^{n} \mu_j \mathrm{d}N_j$$

where there are *n* chemical constituents of the system and permeably connected surrounding subsystems, and where T, S, P, V, N_j, and μ_j, are defined as above.

For a general natural process, there is no immediate term-wise correspondence between equations (2) and (3), because they describe the process in different conceptual frames.

Nevertheless, a conditional correspondence exists. There are three relevant kinds of wall here: purely diathermal, adiabatic, and permeable to matter. If two of those kinds of wall are sealed off, leaving only one that permits transfers of energy, as work, as heat, or with matter, then the remaining permitted terms correspond precisely. If two of the kinds of wall are left unsealed, then energy transfer can be shared between them, so that the two remaining permitted terms do not correspond precisely.

For the special fictive case of quasi-static transfers, there is a simple correspondence. For this, it is supposed that the system has multiple areas of contact with its surroundings. There are pistons that allow adiabatic work, purely diathermal walls, and open connections with surrounding subsystems of completely controllable chemical potential (or equivalent controls for charged species). Then, for a suitable fictive quasi-static transfer, one can write

$\delta Q = T\mathrm{d}S$ and $\delta W = P\mathrm{d}V$ (suitably defined surrounding subsystems, quasi-static transfers of energy).

For fictive quasi-static transfers for which the chemical potentials in the connected surrounding subsystems are suitably controlled, these can be put into equation (3) to yield

$$(4)\ \mathrm{d}U_0 = \delta Q - \delta W + \sum_{j=1}^{n} \mu_j \mathrm{d}N_j \quad \text{(suitably defined surrounding subsystems, quasi-static transfers).}$$

The reference does not actually write equation (4), but what it does write is fully compatible with it. Another helpful account is given by Tschoegl.

There are several other accounts of this, in apparent mutual conflict.

Non-Equilibrium Transfers

The transfer of energy between an open system and a single contiguous subsystem of its surroundings is considered also in non-equilibrium thermodynamics. The problem of definition arises also in this case. It may be allowed that the wall between the system and the subsystem is not only permeable to matter and to internal energy, but also may be movable so as to allow work to be done when the two systems have different pressures. In this case, the transfer of energy as heat is not defined.

Methods for study of non-equilibrium processes mostly deal with spatially continuous flow systems. In this case, the open connection between system and surroundings is usually taken to fully surround the system, so that there are no separate connections impermeable to matter but permeable to heat. Except for the special case mentioned above when there is no actual transfer of matter, which can be treated as if for a closed system, in strictly defined thermodynamic terms, it follows that transfer of energy as heat is not defined. In this sense, there is no such thing as 'heat flow' for a continuous-flow open system. Properly, for closed systems, one speaks of transfer of internal energy as heat, but in general, for open systems, one can speak safely only of transfer of internal energy. A factor here is that there are often cross-effects between distinct transfers, for example that transfer of one substance may cause transfer of another even when the latter has zero chemical potential gradient.

Usually transfer between a system and its surroundings applies to transfer of a state variable, and obeys a balance law, that the amount lost by the donor system is equal to the amount gained by the receptor system. Heat is not a state variable. For his 1947 definition of "heat transfer" for discrete open systems, the author Prigogine carefully explains at some length that his definition of it does not obey a balance law. He describes this as paradoxical.

The situation is clarified by Gyarmati, who shows that his definition of "heat transfer", for continuous-flow systems, really refers not specifically to heat, but rather to transfer of internal energy, as follows. He considers a conceptual small cell in a situation of continuous-flow as a system defined in the so-called Lagrangian way, moving with the local centree of mass. The flow of matter across the boundary is zero when considered as a flow of total mass. Nevertheless if the material constitution is of several chemically distinct components that can diffuse with respect to one another, the system is considered to be open, the diffusive flows of the components being defined with respect to the centree of mass of the system, and balancing one another as to mass transfer. Still there can be a distinction between bulk flow of internal energy and diffusive flow of internal energy in this case, because the internal energy density does not have to be constant per unit mass of material, and allowing for non-conservation of internal energy because of local conversion of kinetic energy of bulk flow to internal energy by viscosity.

Gyarmati shows that his definition of "the heat flow vector" is strictly speaking a definition of flow of internal energy, not specifically of heat, and so it turns out that his use here of the word heat is contrary to the strict thermodynamic definition of heat, though it is more or less compatible with historical custom, that often enough did not clearly distinguish between heat and internal energy; he writes "that this relation must be considered to be the exact definition of the concept of heat flow, fairly loosely used in experimental physics and heat technics." Apparently in a different frame of thinking from that of the above-mentioned paradoxical usage in the earlier sections of the historic 1947 work by Prigogine, about discrete systems, this usage of Gyarmati is consistent with the later sections of the same 1947 work by Prigogine, about continuous-flow systems, which use the term "heat flux" in just this way. This usage is also followed by Glansdorff and Prigogine in their 1971 text about continuous-flow systems. They write: "Again the flow of internal energy may be split into a convection flow $\rho u\mathrm{v}$ and a conduction flow. This conduction flow is by definition the heat flow W. Therefore: $\mathrm{j}[U] = \rho u\mathrm{v} + \mathrm{W}$ where u denotes the [internal] energy per unit mass. [These authors actually use the symbols E and e to denote internal energy but their notation has been changed here to accord with the notation of the present article. These authors actually use the symbol U to refer to total energy, including kinetic energy of bulk flow.]" This usage is followed also by other writers on non-equilibrium thermodynamics such as Lebon, Jou, and Casas-Vásquez, and de Groot and Mazur. This usage is described by Bailyn as stating the non-convective flow of internal energy, and is listed as his definition number 1, according to the first law of thermodynamics. This usage is also followed by workers in the kinetic theory of gases. This is not the *ad hoc* definition of "reduced heat flux" of Haase.

In the case of a flowing system of only one chemical constituent, in the Lagrangian representation, there is no distinction between bulk flow and diffusion of matter. Moreover, the flow of matter is zero into or out of the cell that moves with the local centree of mass. In effect, in this description, one is dealing with a system effectively closed to the transfer of matter. But still one can validly talk of a distinction between bulk flow and diffusive flow of internal energy, the latter driven by a temperature gradient within the flowing material, and being defined with respect to the local centree of mass of the bulk flow. In this case of a virtually closed system, because of the zero matter transfer, as noted above, one can safely distinguish between transfer of energy as work, and transfer of internal energy as heat.

Thermodynamic Free Energy

The thermodynamic free energy is the amount of work that a thermodynamic system can perform. The concept is useful in the thermodynamics of chemical or thermal processes in engineering and science. The free energy is the internal energy of a system minus the amount of energy that cannot be used to perform work. This unusable energy is given by the entropy of a system multiplied by the temperature of the system.

Like the internal energy, the free energy is a thermodynamic state function. Energy is a generalization of free energy.

Overview

Free energy is that portion of any first-law energy that is available to perform thermodynamic work; *i.e.*, work mediated by thermal energy. Free energy is subject to irreversible loss in the course of such work. Since first-law energy is always conserved, it is evident that free energy is an expendable, second-law kind of energy that can perform work within finite amounts of time. Several free energy functions may be formulated based on system criteria. Free energy functions are Legendre transformations of the internal energy. For processes involving a system at constant pressure p and temperature T, the Gibbs free energy is the most useful because, in addition to subsuming any entropy change due merely to heat, it does the same for the $p\mathrm{d}V$ work needed to "make space for additional molecules" produced by various processes. (Hence its utility to solution-phase chemists, including biochemists.) The Helmholtz free energy has a special theoretical importance since it is proportional to the logarithm of the partition function for the canonical ensemble in statistical mechanics. (Hence its utility to physicists; and to gas-phase chemists and engineers, who do not want to ignore $p\mathrm{d}V$ work.)

The historically earlier Helmholtz free energy is defined as $A = U - TS$, where U is the internal energy, T is the absolute temperature, and S is the entropy. Its change is equal to the amount of reversible work done on, or obtainable from, a system at constant T. Thus its appellation "work content", and the designation A from *Arbeit*, the German word for work. Since it makes no reference to any quantities involved in work (such as p and V), the Helmholtz function is completely general: its decrease is the maximum amount of work which can be done *by* a system, and it can increase at most by the amount of work done *on* a system.

The Gibbs free energy is given by $G = H - TS$, where H is the enthalpy. ($H = U + pV$, where p is the pressure and V is the volume.)

Historically, these energy terms have been used inconsistently. In physics, *free energy* most often refers to the Helmholtz free energy, denoted by A, while in chemistry, *free energy* most often refers to the Gibbs free energy. Since both fields use both functions, a compromise has been suggested, using A to denote the Helmholtz function and G for the Gibbs function. While A is preferred by IUPAC, G is sometimes still in use, and the correct free energy function is often implicit in manuscripts and presentations.

Meaning of "Free"

In the 18th and 19th centuries, the theory of heat, i.e., that heat is a form of energy having relation to vibratory motion, was beginning to supplant both the caloric theory, i.e., that heat is a fluid, and the four element theory, in which heat was the lightest of the four elements. In a similar manner, during these years, heat was beginning to be distinguished into different classification categories, such as "free heat", "combined heat", "radiant heat", specific heat, heat capacity, "absolute heat", "latent caloric", "free" or "perceptible" caloric (*calorique sensible*), among others.

In 1780, for example, Laplace and Lavoisier stated: "In general, one can change the first hypothesis into the second by changing the words 'free heat, combined heat, and heat released' into 'vis viva, loss of vis viva, and increase of vis viva.'" In this manner, the total mass of caloric in a body, called *absolute heat*, was regarded as a mixture of two components; the free or perceptible caloric could affect a thermometer, whereas the other component, the latent caloric, could not. The use of the words "latent heat" implied a similarity to latent heat in the more usual sense; it was regarded as chemically bound to the molecules of the body. In the adiabatic compression of a gas, the absolute heat remained constant but the observed rise in temperature implied that some latent caloric had become "free" or perceptible.

During the early 19th century, the concept of perceptible or free caloric began to be referred to as "free heat" or heat set free. In 1824, for example, the French physicist Sadi Carnot, in his famous "Reflections on the Motive Power of Fire", speaks of quantities of heat 'absorbed or set free' in different transformations. In 1882, the German physicist and physiologist Hermann von Helmholtz coined the phrase 'free energy' for the expression $E - TS$, in which the change in F (or G) determines the amount of energy 'free' for work under the given conditions.

Thus, in traditional use, the term "free" was attached to Gibbs free energy, i.e., for systems at constant pressure and temperature, or to Helmholtz free energy, i.e., for systems at constant volume and temperature, to mean 'available in the form of useful work.' With reference to the Gibbs free energy, we add the qualification that it is the energy free for non-volume work.

An increasing number of books and journal articles do not include the attachment "free", referring to G as simply Gibbs energy (and likewise for the Helmholtz energy). This is the result of a 1988 IUPAC meeting to set unified terminologies for the international scientific community, in which the adjective 'free' was supposedly banished. This standard, however, has not yet been universally adopted, and many published articles and books still include the descriptive 'free'.

Application

The experimental usefulness of these functions is restricted to conditions where certain variables (T, and V or *external* p) are held constant, although they also have theoretical importance in deriving Maxwell relations. Work other than $p\mathrm{d}V$ may be added, e.g., for electrochemical cells, or $f \cdot dx$ work in elastic materials and in muscle contraction. Other forms of work which must sometimes be considered are stress-strain, magnetic, as in adiabatic demagnetization used in the approach to absolute zero, and work due to electric polarization. These are described by tensors.

In most cases of interest there are internal degrees of freedom and processes, such as chemical reactions and phase transitions, which create entropy. Even for homogeneous "bulk" materials, the free energy functions depend on the (often suppressed) composition, as do all proper thermodynamic potentials (extensive functions), including the internal energy.

Name	*Symbol*	*Formula*	*Natural variables*
Helmholtz free energy	F	$U - TS$	$T, V, \{N_i\}$
Gibbs free energy	G	$U + pV - TS$	$T, p, \{N_i\}$

N_i is the number of molecules (alternatively, moles) of type i in the system. If these quantities do not appear, it is impossible to describe compositional changes. The differentials for reversible processes are (assuming only pV work):

$$dF = -pdV - SdT + \sum_i \mu_i dN_i$$

$$dG = Vdp - SdT + \sum_i \mu_i dN_i$$

where μ_i is the chemical potential for the i-th component in the system. The second relation is especially useful at constant T and p, conditions which are easy to achieve experimentally, and which approximately characterize living creatures.

$$(dG)_{T,p} = \sum_i \mu_i dN_i$$

Any decrease in the Gibbs function of a system is the upper limit for any isothermal, isobaric work that can be captured in the surroundings, or it may simply be dissipated, appearing as T times a corresponding increase in the entropy of the system and/or its surrounding.

An example is surface free energy, the amount of increase of free energy when the area of surface increases by every unit area.

The path integral Monte Carlo method is a numerical approach for determining the values of free energies, based on quantum dynamical principles.

History

The quantity called "free energy" is a more advanced and accurate replacement for the outdated term *affinity*, which was used by chemists in previous years to describe the *force* that caused chemical reactions. The term affinity, as used in chemical relation, dates back to at least the time of Albertus Magnus in 1250.

From the 1998 textbook *Modern Thermodynamics* by Nobel Laureate and chemistry professor Ilya Prigogine we find: "As motion was explained by the Newtonian concept of force, chemists wanted a similar concept of 'driving force' for chemical change. Why do chemical reactions occur, and why do they stop at certain points? Chemists called the 'force' that caused chemical reactions affinity, but it lacked a clear definition."

During the entire 18th century, the dominant view with regard to heat and light was that put forth by Isaac Newton, called the *Newtonian hypothesis*, which states that light and heat are forms of matter attracted or repelled by other forms of matter, with forces analogous to gravitation or to chemical affinity.

In the 19th century, the French chemist Marcellin Berthelot and the Danish chemist Julius Thomsen had attempted to quantify affinity using heats of reaction. In 1875, after quantifying the heats of reaction for a large number of compounds, Berthelot proposed the *principle of maximum work*, in which all chemical changes occurring without intervention of outside energy tend toward the production of bodies or of a system of bodies which liberate heat.

In addition to this, in 1780 Antoine Lavoisier and Pierre-Simon Laplace laid the foundations of thermochemistry by showing that the heat given out in a reaction is equal to the heat absorbed in the reverse reaction. They also investigated the specific heat and latent heat of a number of substances, and amounts of heat given out in combustion. In a similar manner, in 1840 Swiss chemist Germain Hess formulated the principle that the evolution of heat in a reaction is the same whether the process is accomplished in one-step process or in a number of stages. This is known as Hess' law. With the advent of the mechanical theory of heat in the early 19th century, Hess's law came to be viewed as a consequence of the law of conservation of energy.

Based on these and other ideas, Berthelot and Thomsen, as well as others, considered the heat given out in the formation of a compound as a measure of the affinity, or the work done by the chemical forces. This view, however, was not entirely correct. In 1847, the English physicist James Joule showed that he could raise the temperature of water by turning a paddle wheel in it, thus showing that heat and mechanical work were equivalent or proportional to each other, i.e., approximately, $dW \propto dQ$. This statement came to be known as the mechanical equivalent of heat and was a precursory form of the first law of thermodynamics.

By 1865, the German physicist Rudolf Clausius had shown that this equivalence principle needed amendment. That is, one can use the heat derived from a combustion reaction in a coal furnace to boil water, and use this heat to vaporize steam, and then use the enhanced high-pressure energy of the vaporized steam to push a piston. Thus, we might naively reason that one can entirely convert the initial combustion heat of the chemical reaction into the work of pushing the piston. Clausius showed, however, that we must take into account the work that the molecules of the working body, i.e., the water molecules in the cylinder, do on each other as they pass or transform from one step of or state of the engine cycle to the next, e.g., from (P_1, V_1) to (P_2, V_2). Clausius originally called this the "transformation content" of the body, and then later changed the name to entropy. Thus, the heat used to transform the working body of molecules from one state to the next cannot be used to do external work, e.g., to push the piston. Clausius defined this *transformation heat* as $dQ = TdS$.

In 1873, Willard Gibbs published *A Method of Geometrical Representation of the Thermodynamic Properties of Substances by Means of Surfaces*, in which he introduced the preliminary outline of the principles of his new equation able to predict or estimate the tendencies of various natural processes to ensue when bodies or systems are brought into contact. By studying the interactions of homogeneous substances in contact, i.e., bodies, being in composition part solid, part liquid, and part vapor, and by using a three-dimensional volume-entropy-internal energy graph, Gibbs was able to determine three states of equilibrium, i.e., "necessarily stable", "neutral", and "unstable", and whether or not changes will ensue. In 1876, Gibbs built on this framework by introducing the concept of chemical potential so to take into account chemical reactions and states of bodies that are chemically different from each other. In his own words, to summarize his results in 1873, Gibbs states:

> If we wish to express in a single equation the necessary and sufficient condition of thermodynamic equilibrium for a substance when surrounded by a medium of constant pressure p and temperature T, this equation may be written:$\delta(\varepsilon - T\eta + pv) = 0$when δ refers to the variation produced by any variations in the state of the parts of the body, and (when different parts of the body are in different states) in the proportion in which the body is divided between the different states. The condition of stable equilibrium is that the value of the expression in the parenthesis shall be a minimum.

In this description, as used by Gibbs, ε refers to the internal energy of the body, η refers to the entropy of the body, and v is the volume of the body.

Hence, in 1882, after the introduction of these arguments by Clausius and Gibbs, the German scientist Hermann von Helmholtz stated, in opposition to Berthelot and Thomas' hypothesis that chemical affinity is a measure of the heat of reaction of chemical reaction as based on the principle of maximal work, that affinity is not the heat given out in the formation of a compound but rather it is the largest quantity of work which can be gained when the reaction is carried out in a reversible manner, e.g., electrical work in a reversible cell. The maximum work is thus regarded as the diminution of the free, or available, energy of the system (*Gibbs free energy* G at T = constant, P = constant or *Helmholtz free energy* F at T = constant, V = constant), whilst the heat given out is usually a measure of the diminution of the total energy of the system (Internal energy). Thus, G or F is the amount of energy "free" for work under the given conditions.

Up until this point, the general view had been such that: "all chemical reactions drive the system to a state of equilibrium in which the affinities of the reactions vanish". Over the next 60 years, the term affinity came to be replaced with the term free energy. According to chemistry historian Henry Leicester, the influential 1923 textbook *Thermodynamics and the Free Energy of Chemical Reactions* by Gilbert N. Lewis and Merle Randall led to the replacement of the term "affinity" by the term "free energy" in much of the English-speaking world.

Gas Turbine

A gas turbine, also called a combustion turbine, is a type of internal combustion engine. It has an upstream rotating compressor coupled to a downstream turbine, and a combustion chamber in-between.

The basic operation of the gas turbine is similar to that of the steam power plant except that air is used instead of water. Fresh atmospheric air flows through a compressor that brings it to higher pressure. Energy is then added by spraying fuel into the air and igniting it so the combustion generates a high-temperature flow. This high-temperature high-pressure gas enters a turbine, where it expands down to the exhaust pressure, producing a shaft work output in the process. The turbine shaft work is used to drive the compressor and other devices such as an electric generator that may be coupled to the shaft. The energy that is not used for shaft work comes out in the exhaust gases, so these have either a high temperature or a high velocity. The purpose of the gas turbine determines the design so that the most desirable energy form is maximized. Gas turbines are used to power aircraft, trains, ships, electrical generators, or even tanks.

Theory of Operation

In an ideal gas turbine, gases undergo three thermodynamic processes: an isentropic compression, an isobaric (constant pressure) combustion and an isentropic expansion. Together, these make up the Brayton cycle.

In a practical gas turbine, mechanical energy is irreversibly transformed into heat when gases are compressed (in either a centrifugal or axial compressor), due to internal friction and turbulence.

Passage through the combustion chamber, where heat is added and the specific volume of the gases increases, is accompanied by a slight loss in pressure. During expansion amidst the stator and rotor blades of the turbine, irreversible energy transformation once again occurs.

If the device has been designed to power a shaft as with an industrial generator or a turboprop, the exit pressure will be as close to the entry pressure as possible. In practice it is necessary that some pressure remains at the outlet in order to fully expel the exhaust gases. In the case of a jet engine only enough pressure and energy is extracted from the flow to drive the compressor and other components. The remaining high pressure gases are accelerated to provide a jet that can, for example, be used to propel an aircraft.

As with all cyclic heat engines, higher combustion temperatures can allow for greater efficiencies. However, temperatures are limited by ability of the steel, nickel, ceramic, or other materials that make up the engine to withstand high temperatures and stresses. To combat this many turbines feature complex blade cooling systems.

As a general rule, the smaller the engine, the higher the rotation rate of the shaft(s) must be to maintain tip speed. Blade-tip speed determines the maximum pressure ratios that can be obtained by the turbine and the compressor. This, in turn, limits the maximum power and efficiency that can be obtained by the engine. In order for tip speed to remain constant, if the diameter of a rotor is reduced by half, the rotational speed must double. For example, large jet engines operate around 10,000 rpm, while micro turbines spin as fast as 500,000 rpm.

Mechanically, gas turbines can be considerably less complex than internal combustion piston engines. Simple turbines might have one moving part: the shaft/compressor/turbine/alternative-rotor assembly, not counting the fuel system. However, the required precision manufacturing for components and temperature resistant alloys necessary for high efficiency often make the construction of a simple turbine more complicated than piston engines.

More sophisticated turbines (such as those found in modern jet engines) may have multiple shafts (spools), hundreds of turbine blades, movable stator blades, and a vast system of complex piping, combustors and heat exchangers.

Thrust bearings and journal bearings are a critical part of design. Traditionally, they have been hydrodynamic oil bearings, or oil-cooled ball bearings. These bearings are being surpassed by foil bearings, which have been successfully used in micro turbines and auxiliary power units.

Combined Cycle Plant for Power Generation

The process for converting the energy in a fuel into electric power involves the creation of mechanical work, which is then transformed into electric power by a generator. Depending on the fuel type and thermodynamic process, the overall efficiency of this conversion can be as low as 30 percent. This means that two-thirds of the latent energy of the fuel ends up wasted. For example, steam electric power plants which utilise boilers to combust a fossil fuel average 33 percent efficiency. Simple cycle gas turbine (GTs) plants average just under 30 percent efficiency on natural gas, and around 25 percent on fuel oil. Much of this wasted energy ends up as thermal energy in the hot exhaust gases from the combustion process.

To increase the overall efficiency of electric power plants, multiple processes can be combined to recover and utilise the residual heat energy in hot exhaust gases. In combined cycle mode, power plants can achieve electrical efficiencies up to 60 percent. The term "combined cycle" refers to the combining of multiple thermodynamic cycles to generate power. Combined cycle operation employs a heat recovery steam generator (HRSG) that captures heat from high temperature exhaust gases to produce steam, which is then supplied to a steam turbine to generate additional electric power. The process for creating steam to produce work using a steam turbine is based on the Rankine cycle.

The most common type of combined cycle power plant utilises gas turbines and is called a combined cycle gas turbine (CCGT) plant. Because gas turbines have low efficiency in simple cycle operation, the output produced by the steam turbine accounts for about half of the CCGT plant output. There are many different configurations for CCGT power plants, but typically each GT has its own associated HRSG, and multiple HRSGs supply steam to one or more steam turbines. For example, at a plant in a 2x1 configuration, two GT/HRSG trains supply to one steam turbine; likewise there can be 1x1, 3x1 or 4x1 arrangements. The steam turbine is sized to the number and capacity of supplying GTs/HRSGs.

Combined Cycle Principles of Operation

The HRSG is basically a heat exchanger, or rather a series of heat exchangers. It is also called a boiler, as it creates steam for the steam turbine by passing the hot exhaust gas flow from a gas turbine or combustion engine through banks of heat exchanger tubes. The HRSG can rely on natural circulation or utilise forced circulation using pumps. As the hot exhaust gases flow past the heat exchanger tubes in which hot water circulates, heat is absorbed causing the creation of steam in the tubes. The tubes are arranged in sections, or modules, each serving a different function in the production of dry superheated steam. These modules are referred to as economizers, evaporators, superheaters/ reheaters and preheaters.

The economizer is a heat exchanger that preheats the water to approach the saturation temperature (boiling point), which is supplied to a thick-walled steam drum. The drum is located adjacent to finned evaporator tubes that circulate heated water. As the hot exhaust gases flow past the evaporator tubes, heat is absorbed causing the creation of steam in the tubes. The steam-water mixture in the tubes enters the steam drum where steam is separated from the hot water using moisture separators and cyclones. The separated water is recirculated to the evaporator tubes. Steam drums also serve storage and water treatment functions. An alternative design to steam drums is a once-through HRSG, which replaces the steam drum with thin-walled components that are better suited to handle changes in exhaust gas temperatures and steam pressures during frequent starts and stops. In some designs, duct burners are used to add heat to the exhaust gas stream and boost steam production; they can be used to produce steam even if there is insufficient exhaust gas flow.

Saturated steam from the steam drums or once-through system is sent to the superheater to produce dry steam which is required for the steam turbine. Preheaters are located at the coolest end of the HRSG gas path and absorb energy to preheat heat exchanger liquids, such as water/glycol mixtures, thus extracting the most economically viable amount of heat from exhaust gases.

The superheated steam produced by the HRSG is supply to the steam turbine where it expands through the turbine blades, imparting rotation to the turbine shaft. The energy delivered to the generator drive shaft is converted into electricity. After exiting the steam turbine, the steam is sent to a condenser which routes the condensed water back to the HRSG.

CCGT Design Considerations

Designs and configurations for HRSGs and steam turbines depend on the exhaust gas characteristics, steam requirements, and expected power plant operations. Because the exhaust gases from a gas turbine can reach 600°C, HRSGs for GTs may produce steam at multiple pressure levels to optimize energy recovery; thus they often have three sets of heat exchanger modules – one for high pressure (HP) steam, one for intermediate pressure (IP) steam, and one for low pressure (LP) steam. The high pressure steam in a large CCGT plant can reach 40 – 110 bar. With a multiple-pressure HRSG, the steam turbine will typically have multiple steam admission points. In a three-stage steam turbine, HP, IP and LP steam produced by the HRSG is fed into the turbine at different points.

The HRSGs present operational constraints on the CCGT power plant. As the HRSGs are located directly downstream of the gas turbines, changes in temperature and pressure of the exhaust gases cause thermal and mechanical stress. When CCGT power plants are used for load-following operation, characterized by frequent starts and stops or operating at part-load to meet fluctuating electric demand, this cycling can cause thermal stress and eventual damage in some components of the HRSG. The HP steam drum and superheater headers are more prone to reduced mechanical life because they are subjected to the highest exhaust gas temperatures. Important design and operating considerations are the gas and steam temperatures that the module materials can withstand; mechanical stability for turbulent exhaust flow; corrosion of HRSG tubes; and steam pressures that may necessitate thicker-walled drums. To control the rate of pressure and temperature increase in HRSG components, bypass systems can be used to divert some of the GT exhaust gases from entering the HRSG during startup.

The HRSG takes longer to warm up from cold conditions than from hot conditions. As a result, the amount of time elapsed since last shutdown influences startup time. When gas turbines are ramped to load quickly, the temperature and flow in the HRSG may not yet have achieved conditions to produce steam, which causes metal overheating since there is no cooling steam flow. In 1x1 configurations, the operation of the steam turbine is directly coupled to the GT/HRSG operation, limiting the rate at which the power plant can be ramped to load. Steam conditions acceptable for the steam turbine are dictated by thermal limits of the rotor, blade, and casing design.

Control equipment for nitrogen oxides (NOx) and carbon monoxide (CO) emissions are integrated into the HRSG. As these systems operate efficiently over a narrow range of gas temperatures, they are often installed between evaporator modules.

Flexible Combined Cycle: The Flexicycle Power Plant

The Flexicycle power plant is a combined cycle power plant with unique characteristics based on Wärtsilä gas or dual-fuel reciprocating engines. Because reciprocating engines convert more of the fuel energy into mechanical work, they have higher simple cycle efficiencies, averaging near 50 percent. The exhaust gases from reciprocating internal combustion engines are around 360°C, much lower temperature than GT exhaust. Due to the lower exhaust gas temperatures, HRSGs designed for reciprocating engine power plants are much simpler in design, creating steam at one pressure level – approximately 15 bar. The steam turbine process adds approximately 20% to the efficiency of the Flexicycle power plant.

In a Flexicycle power plant, each reciprocating engine generator set has an associated HRSG. Bypass valves are used to control the admission of steam to the steam turbine when an engine set is not operating. One engine can be used to preheat all the HRSG exhaust gas boilers with steam to keep the HRSGs hot and enable fast starting. Flexicycle power plants combine the advantages of high efficiency in simple cycle and the modularity of multiple engines supplying the steam turbine. The steam turbine can be run with only 25 percent of the engines at full load, or 50 percent of the engines at half load. For a 12-engine power plant of around 200 megawatts (MW), this means only three of the engines need to be operating to produce enough steam to run the steam turbine. The result is a very efficient power plant that retains the operational agility of a power plant based on simple-cycle engines.

Combustion of Fossil Fuels

The combustion of all fossil fuels follows a very similar reaction: Fuel (any hydrocarbon source) plus oxygen yields carbon dioxide and water and energy.

A simple combustion reaction is given for methane. The combustion of methane means that it is possible to burn it. Chemically, this combustion process consists of a reaction between methane and oxygen in the air. When this reaction takes place, the result is carbon dioxide (CO_2), water (H_2O), and a great deal of energy. The following reaction represents the combustion of methane:

$$CH_4[g] + 2\ O_2[g] \rightarrow CO_2[g] + 2\ H_2O[g] + \text{energy}$$

One molecule of methane, (the [g] referred to above means it is gaseous form), combined with two oxygen molecules, react to form a carbon dioxide molecule, and two water molecules usually given off as steam or water vapor during the reaction and energy.

Natural gas is the cleanest burning fossil fuel. Coal and oil, the other fossil fuels, are more chemically complicated than natural gas, and when combusted, release a variety of potentially harmful air pollutants. Burning methane releases only carbon dioxide and water. Since natural gas is mostly methane, the combustion of natural gas releases fewer byproducts than other fossil fuels.

Fluidized Bed

A fluidized bed is formed when a quantity of a solid particulate substance (usually present in a holding vessel) is placed under appropriate conditions to cause a solid/fluid mixture to behave as a fluid. This is usually achieved by the introduction of pressurized fluid through the particulate medium. This results in the medium then having many properties and characteristics of normal fluids, such as the ability to free-flow under gravity, or to be pumped using fluid type technologies.

The resulting phenomenon is called fluidization. Fluidized beds are used for several purposes, such as fluidized bed reactors (types of chemical reactors), fluid catalytic cracking, fluidized bed combustion, heat or mass transfer or interface modification, such as applying a coating onto solid items. This technique is also becoming more common in aquaculture for the production of shellfish in integrated multi-trophic aquaculture systems.

Properties of Fluidized Beds

A fluidized bed consists of fluid-solid mixture that exhibits fluid-like properties. As such, the upper surface of the bed is relatively horizontal, which is analogous to hydrostatic behaviour. The bed can be considered to be a heterogeneous mixture of fluid and solid that can be represented by a single bulk density.

Furthermore, an object with a higher density than the bed will sink, whereas an object with a lower density than the bed will float, thus the bed can be considered to exhibit the fluid behaviour expected of Archimedes' principle. As the "density", (actually the solid volume fraction of the suspension), of the bed can be altered by changing the fluid fraction, objects with different densities comparative to the bed can, by altering either the fluid or solid fraction, be caused to sink or float.

In fluidized beds, the contact of the solid particles with the fluidization medium (a gas or a liquid) is greatly enhanced when compared to packed beds. This behaviour in fluidized combustion beds enables good thermal transport inside the system and good heat transfer between the bed and its container. Similarly to the good heat transfer, which enables thermal uniformity analogous to that of a well mixed gas, the bed can have a significant heat-capacity whilst maintaining a homogeneous temperature field.

Fluidized Bed Types

Bed types can be coarsely classified by their flow behaviour, including:

- Stationary or bubbling bed is the classical approach where the gas at low velocities is used and fluidization of the solids is relatively stationary, with some fine particles being entrained.
- Circulating fluidized beds (CFB), where gases are at a higher

velocity sufficient to suspend the particle bed, due to a larger kinetic energy of the fluid. As such the surface of the bed is less smooth and larger particles can be entrained from the bed than for stationary beds. Entrained particles are recirculated via an external loop back into the reactor bed. Depending on the process, the particles may be classified by a cyclone separator and separated from or returned to the bed, based upon particle cut size.

- Vibratory fluidized beds are similar to stationary beds, but add a mechanical vibration to further excite the particles for increased entrainment.
- Transport or flash reactor (FR). At velocities higher than CFB, particles approach the velocity of the gas. Slip velocity between gas and solid is significantly reduced at the cost of less homogeneous heat distribution.
- Annular fluidized bed (AFB). A large nozzle at the centree of a bubble bed introduces gas as high velocity achieving the rapid mixing zone above the surrounding bed comparable to that found in the external loop of a CFB.

Carbon Sequestration

Carbon sequestration is the process of capture and long-term storage of atmospheric carbon dioxide (CO_2) and may refer specifically to:

- "The process of removing carbon from the atmosphere and depositing it in a reservoir." When carried out deliberately, this may also be referred to as carbon dioxide removal, which is a form of geoengineering.
- The process of carbon capture and storage, where carbon dioxide is removed from flue gases, such as on power stations, before being stored in underground reservoirs.
- Natural biogeochemical cycling of carbon between the atmosphere and reservoirs, such as by chemical weathering of rocks.

Carbon sequestration describes long-term storage of carbon dioxide or other forms of carbon to either mitigate or defer global warming and avoid dangerous climate change. It has been proposed as a way to slow the atmospheric and marine accumulation of greenhouse gases, which are released by burning fossil fuels.

Carbon dioxide is naturally captured from the atmosphere through biological, chemical or physical processes. Some anthropogenic sequestration techniques exploit these natural processes, while some use entirely artificial processes.

Carbon dioxide may be captured as a pure by-product in processes related to petroleum refining or from flue gases from power generation.

CO_2 sequestration includes the storage part of carbon capture and storage, which refers to large-scale, permanent artificial capture and sequestration of industrially produced CO_2 using subsurface saline aquifers, reservoirs, ocean water, aging oil fields, or other carbon sinks.

Geothermal Energy

Geothermal energy is the heat from the Earth. It's clean and sustainable. Resources of geothermal energy range from the shallow ground to hot water and hot rock found a few miles beneath the Earth's surface, and down even deeper to the extremely high temperatures of molten rock called magma. Almost everywhere, the shallow ground or upper 10 feet of the Earth's surface maintains a nearly constant temperature between 50° and 60°F (10° and 16°C). Geothermal heat pumps can tap into this resource to heat and cool buildings. A geothermal heat pump system consists of a heat pump, an air delivery system (ductwork), and a heat exchanger-a system of pipes buried in the shallow ground near the building. In the winter, the heat pump removes heat from the heat exchanger and pumps it into the indoor air delivery system. In the summer, the process is reversed, and the heat pump moves heat from the indoor air into the heat exchanger. The heat removed from the indoor air during the summer can also be used to provide a free source of hot water.

In the United States, most geothermal reservoirs of hot water are located in the western states, Alaska, and Hawaii. Wells can be drilled into underground reservoirs for the generation of electricity. Some geothermal power plants use the steam from a reservoir to power a turbine/ generator, while others use the hot water to boil a working fluid that vaporizes and then turns a turbine. Hot water near the surface of Earth can be used directly for heat. Direct-use applications include heating buildings, growing plants in greenhouses, drying crops, heating water at fish farms, and several industrial processes such as pasteurizing milk.

Hot dry rock resources occur at depths of 3 to 5 miles everywhere beneath the Earth's surface and at lesser depths in certain areas. Access to these resources involves injecting cold water down one well, circulating it through hot fractured rock, and drawing off the heated water from another well. Currently, there are no commercial applications of this technology. Existing technology also does not yet allow recovery of heat directly from magma, the very deep and most powerful resource of geothermal energy.

Chapter 3

Sources of Energy

Hydropower

Hydro-power or water power is power derived from the energy of falling water and running water, which may be harnessed for useful purposes. Since ancient times, hydro-power has been used for irrigation and the operation of various mechanical devices, such as watermills, sawmills, textile mills, dock cranes, domestic lifts, power houses and paint making.Since the early 20th century, the term has been used almost exclusively in conjunction with the modern development of hydro-electric power, which allowed use of distant energy sources. Another method used to transmit energy is by using a trompe, which produces compressed air from falling water. Compressed air could then be piped to power other machinery at a distance from the waterfall. Hydro power is a renewable energy source.

Water's power is manifested in hydrology, by the forces of water on the riverbed and banks of a river. When a river is in flood, it is at its most powerful, and moves the greatest amount of sediment. This higher force results in the removal of sediment and other material from the riverbed and banks of the river, locally causing erosion, transport and, with lower flow, sedimentation downstream.

Hydropower Types

Hydropower is used primarily to generate electricity. Broad categories include:

- Conventional hydroelectric, referring to hydroelectric dams.
- Run-of-the-river hydroelectricity, which captures the kinetic energy in rivers or streams, without the use of dams.

- Small hydro projects are 10 megawatts or less and often have no artificial reservoirs.

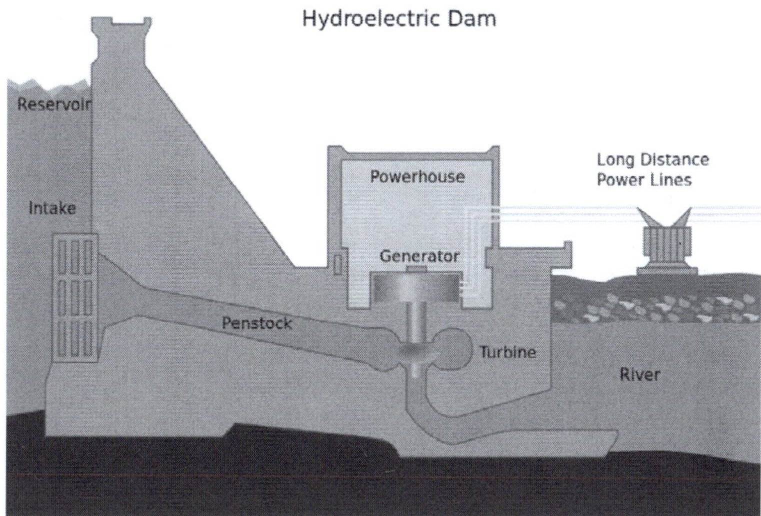

Figure: *A conventional dammed-hydro facility (hydroelectric dam) is the most common type of hydroelectric power generation.*

- Micro hydro projects provide a few kilowatts to a few hundred kilowatts to isolated homes, villages, or small industries.
- Conduit hydroelectricity projects utilise water which has already been diverted for use elsewhere; in a municipal water system for example.
- Pumped-storage hydroelectricity stores water pumped during periods of low demand to be released for generation when demand is high.

Calculating the Amount of Available Power

A hydropower resource can be evaluated by its available power. Power is a function of the hydraulic head and rate of fluid flow. The head is the energy per unit weight (or unit mass) of water. The static head is proportional to the difference in height through which the water falls. Dynamic head is related to the velocity of moving water. Each unit of water can do an amount of work equal to its weight times the head.

The power available from falling water can be calculated from the flow rate and density of water, the height of fall, and the local acceleration due to gravity. In SI units, the power is:

$$P = \eta \rho Q g h$$

where

- P is power in watts
- η is the dimensionless efficiency of the turbine
- ρ is the density of water in kilograms per cubic metre
- Q is the flow in cubic metres per second
- g is the acceleration due to gravity
- h is the height difference between inlet and outlet

To illustrate, power is calculated for a turbine that is 85% efficient, with water at 1000 kg/cubic metre(62.5 pounds/cubic foot) and a flow rate of 80 cubic-meters/second(2800 cubic-feet/second), gravity of 9.81 metres per second squared and with a net head of 145 m (480 ft).

In SI units:

Power (W) $= 0.85 \times 1000 \times 80 \times 9.81 \times 145$ which gives 97 MW

In English units, the density is given in pounds per cubic foot so acceleration due to gravity is inherent in the unit of weight. A conversion factor is required to change from foot lbs/second to kilowatts:

Power (W) $= 0.85 \times 62.5 \times 2800 \times 480 \times 1.356$ which gives 97 MW (130,000 horsepower)

Operators of hydroelectric stations will compare the total electrical energy produced with the theoretical potential energy of the water passing through the turbine to calculate efficiency. Procedures and definitions for calculation of efficiency are given in test codes such as ASME PTC 18 and IEC 60041. Field testing of turbines is used to validate the manufacturer's guaranteed efficiency. Detailed calculation of the efficiency of a hydropower turbine will account for the head lost due to flow friction in the power canal or penstock, rise in tail water level due to flow, the location of the station and effect of varying gravity, the temperature and barometric pressure of the air, the density of the water at ambient temperature, and the altitudes above sea level of the forebay and tailbay. For precise calculations, errors due to rounding and the number of significant digits of constants must be considered.

Some hydropower systems such as water wheels can draw power from the flow of a body of water without necessarily changing its height. In this case, the available power is the kinetic energy of the flowing water. Over-shot water wheels can efficiently capture both types of energy.

The water flow in a stream can vary widely from season to season. Development of a hydropower site requires analysis of flow records, sometimes spanning decades, to assess the reliable annual energy supply. Dams and reservoirs provide a more dependable source of power by smoothing seasonal changes in water flow. However reservoirs have significant environmental impact, as does alteration of naturally occurring stream flow. The design of dams must also account for the worst-case, "probable maximum flood" that can be expected at the site; a spillway is often included to bypass flood flows around the dam. A computer model of the hydraulic basin and rainfall and snowfall records are used to predict the maximum flood.

Hydroelectricity

Hydroelectricity is the term referring to electricity generated by hydropower; the production of electrical power through the use of the gravitational force of falling or flowing water. It is the most widely used form of renewable energy, accounting for 16 percent of global electricity generation – 3,427 terawatt-hours of electricity production in 2010, and is expected to increase about 3.1% each year for the next 25 years.

Hydropower is produced in 150 countries, with the Asia-Pacific region generating 32 percent of global hydropower in 2010. China is the largest hydroelectricity producer, with 721 terawatt-hours of production in 2010, representing around 17 percent of domestic electricity use. There are now four hydroelectricity stations larger than 10 GW: the Three Gorges Dam and Xiluodu Dam in China, Itaipu Dam across the Brazil/Paraguay border, and Guri Dam in Venezuela.

The cost of hydroelectricity is relatively low, making it a competitive source of renewable electricity. The average cost of electricity from a hydro station larger than 10 megawatts is 3 to 5 U.S. cents per kilowatt-hour. It is also a flexible source of electricity since the amount produced by the station can be changed up or down very quickly to adapt to changing energy demands. However, damming interrupts the flow of rivers and can harm local ecosystems, and building large dams and reservoirs often involves displacing people and wildlife. Once a hydroelectric complex is constructed, the project produces no direct waste, and has a considerably lower output level of the greenhouse gas carbon dioxide (CO_2) than fossil fuel powered energy plants.

Generating Methods

Conventional (Dams): Most hydroelectric power comes from the potential energy of dammed water driving a water turbine and generator. The power extracted from the water depends on the volume and on the difference in height between the source and the water's outflow. This height difference is called the head. The amount of potential energy in water is proportional to the head. A large pipe (the "penstock") delivers water to the turbine.

Pumped-Storage

This method produces electricity to supply high peak demands by moving water between reservoirs at different elevations. At times of low electrical demand, excess generation capacity is used to pump water into the higher reservoir. When there is higher demand, water is released back into the lower reservoir through a turbine. Pumped-storage schemes currently provide the most commercially important means of large-scale grid energy storage and improve the daily capacity factor of the generation system. Pumped storage is not an energy source, and appears as a negative number in listings.

Run of the River

Run of the river hydroelectric stations are those with small or no reservoir capacity, so that the water coming from upstream must be used for generation at that moment, or must be allowed to bypass the dam. In the United States, run of the river hydropower could potentially provide 60,000 MW (about 13.7% of total use in 2011 if continuously available).

Tide

A tidal power station makes use of the daily rise and fall of ocean water due to tides; such sources are highly predictable, and if conditions permit construction of reservoirs, can also be dispatchable to generate power during high demand periods. Less common types of hydro schemes use water's kinetic energy or undammed sources such as undershot waterwheels. Tidal power is viable in a relatively small number of locations around the world. In Great Britain, there are eight sites that could be developed, which have the potential to generate 20% of the electricity used in 2012.

Calculating Available Power

A simple formula for approximating electric power production at a hydroelectric station is: $P = \rho hrgk$, where

- P is Power in watts,
- ρ is the density of water (~1000 kg/m^3),
- h is height in meters,
- r is flow rate in cubic meters per second,
- g is acceleration due to gravity of 9.8 m/s^2,

- k is a coefficient of efficiency ranging from 0 to 1. Efficiency is often higher (that is, closer to 1) with larger and more modern turbines.

Annual electric energy production depends on the available water supply. In some installations, the water flow rate can vary by a factor of 10:1 over the course of a year.

Water Turbine

A water turbine is a rotary engine that takes energy from moving water.

Water turbines were developed in the 19th century and were widely used for industrial power prior to electrical grids. Now they are mostly used for electric power generation. Water turbines are mostly found in dams to generate electric power from water kinetic energy.

History

Water wheels have been used for hundreds of years for industrial power. Their main shortcoming is size, which limits the flow rate and head that can be harnessed. The migration from water wheels to modern turbines took about one hundred years. Development occurred during the Industrial revolution, using scientific principles and methods. They also made extensive use of new materials and manufacturing methods developed at the time.

Swirl

The word turbine was introduced by the French engineer Claude Burdin in the early 19th century and is derived from the Latin word for "whirling" or a "vortex". The main difference between early water turbines and water wheels is a swirl component of the water which passes energy to a spinning rotor. This additional component of motion allowed the turbine to be smaller than a water wheel of the same power. They could process more water by spinning faster and could harness much greater heads. (Later, impulse turbines were developed which didn't use swirl).

Time Line

The earliest known water turbines date to the Roman Empire. Two helix-turbine mill sites of almost identical design were found at Chemtou and Testour, modern-day Tunisia, dating to the late 3rd or early 4th century AD. The horizontal water wheel with angled blades was installed at the bottom of a water-filled, circular shaft. The water from the mill-race entered the pit tangentially, creating a swirling water column which made the fully submerged wheel act like a true turbine.

Johann Segner developed a reactive water turbine (Segner wheel) in the mid-18th century in Kingdom of Hungary. It had a horizontal axis and was a precursor to modern water turbines. It is a very simple machine that is still produced today for use in small hydro sites. Segner worked with Euler on some of the early mathematical theories of turbine design. In the 18th century, a Dr. Barker invented a similar reaction hydraulic turbine that became popular as a lecture-hall demonstration. The only known surviving example of this type of engine used in power production, dating from 1851, is found at Hacienda Buena Vista in Ponce, Puerto Rico.

In 1820, Jean-Victor Poncelet developed an inward-flow turbine.

In 1826, Benoit Fourneyron developed an outward-flow turbine. This was an efficient machine (~80%) that sent water through a runner with blades curved in one dimension. The stationary outlet also had curved guides.

In 1844, Uriah A. Boyden developed an outward flow turbine that improved on the performance of the Fourneyron turbine. Its runner shape was similar to that of a Francis turbine.

In 1849, James B. Francis improved the inward flow reaction turbine to over 90% efficiency. He also conducted sophisticated tests and developed engineering methods for water turbine design. The Francis turbine, named for him, is the first modern water turbine. It is still the most widely used water turbine in the world today. The Francis turbine is also called a radial flow turbine, since water flows from the outer circumference towards the centre of runner.

Inward flow water turbines have a better mechanical arrangement and all modern reaction water turbines are of this design. As the water swirls inward, it accelerates, and transfers energy to the runner. Water pressure decreases to atmospheric, or in some cases subatmospheric, as the water passes through the turbine blades and loses energy.

Around 1890, the modern fluid bearing was invented, now universally used to support heavy water turbine spindles. As of 2002, fluid bearings appear to have a mean time between failures of more than 1300 years.

Around 1913, Viktor Kaplan created the Kaplan turbine, a propeller-type machine. It was an evolution of the Francis turbine but revolutionized the ability to develop low-head hydro sites.

New Concept

All common water machines until the late 19th century (including water wheels) were basically reaction machines; water *pressure* head acted on the machine and produced work. A reaction turbine needs to fully contain the water during energy transfer.

In 1866, California millwright Samuel Knight invented a machine that took the impulse system to a new level. Inspired by the high pressure jet systems used in hydraulic mining in the gold fields, Knight developed a bucketed wheel which captured the energy of a free jet, which had converted a high head (hundreds of vertical feet in a pipe or penstock) of water to kinetic energy. This is called an impulse or tangential turbine. The water's velocity, roughly twice the velocity of the bucket periphery, does a u-turn in the bucket and drops out of the runner at low velocity.

In 1879, Lester Pelton, experimenting with a Knight Wheel, developed a Pelton wheel (double bucket design), which exhausted the water to the side, eliminating some energy loss of the Knight wheel which exhausted some water back against the centree of the wheel. In about 1895, William Doble improved on Pelton's half-cylindrical bucket form with an elliptical bucket that included a cut in it to allow the jet a cleaner bucket entry. This is the modern form of the Pelton turbine which today achieves up to 92% efficiency. Pelton had been quite an effective promoter of his design and although Doble took over the Pelton company he did not change the name to Doble because it had brand name recognition.

Turgo and cross-flow turbines were later impulse designs.

Theory of Operation

Flowing water is directed on to the blades of a turbine runner, creating a force on the blades. Since the runner is spinning, the force acts through a distance (force acting through a distance is the definition of work). In this way, energy is transferred from the water flow to the turbine.

Water turbines are divided into two groups; reaction turbines and impulse turbines.

The precise shape of water turbine blades is a function of the supply pressure of water, and the type of impeller selected.

Reaction Turbines

Reaction turbines are acted on by water, which changes pressure as it moves through the turbine and gives up its energy. They must be encased to contain the water pressure (or suction), or they must be fully submerged in the water flow.

Newton's third law describes the transfer of energy for reaction turbines.

Most water turbines in use are reaction turbines and are used in low (<30 m or 100 ft) and medium (30–300 m or 100–1,000 ft) head applications. In reaction turbine pressure drop occurs in both fixed and moving blades. It is largely used in dam and large power plants

Impulse Turbines

Impulse turbines change the velocity of a water jet. The jet pushes on the turbine's curved blades which changes the direction of the flow. The resulting change in momentum (impulse) causes a force on the turbine blades. Since the turbine is spinning, the force acts through a distance (work) and the diverted water flow is left with diminished energy. An impulse turbine is one which the pressure of the fluid flowing over the rotor blades is constant and all the work output is due to the change in kinetic energy of the fluid.

Prior to hitting the turbine blades, the water's pressure (potential energy) is converted to kinetic energy by a nozzle and focused on the turbine. No pressure change occurs at the turbine blades, and the turbine doesn't require a housing for operation.

Newton's second law describes the transfer of energy for impulse turbines. Impulse turbines are often used in very high (>300m/1000 ft) head applications.

Power

The power available in a stream of water is;

$$P = \eta \cdot \rho \cdot g \cdot h \cdot \dot{q}$$

where:

- P = power (J/s or watts)
- η = turbine efficiency
- ρ = density of water (kg/m^3)
- g = acceleration of gravity (9.81 m/s^2)
- h = head (m). For still water, this is the difference in height between the inlet and outlet surfaces. Moving water has an additional component added to account for the kinetic energy of the flow. The total head equals the *pressure head* plus *velocity head*.
- $\dot{q}$ = flow rate (m^3/s)

Pumped-Storage Hydroelectricity

Some water turbines are designed for pumped-storage hydroelectricity. They can reverse flow and operate as a pump to fill a high reservoir during off-peak electrical hours, and then revert to a water turbine for power generation during peak electrical demand. This type of turbine is usually a Deriaz or Francis turbine in design.

Efficiency

Large modern water turbines operate at mechanical efficiencies greater than 90%.

Design and Application

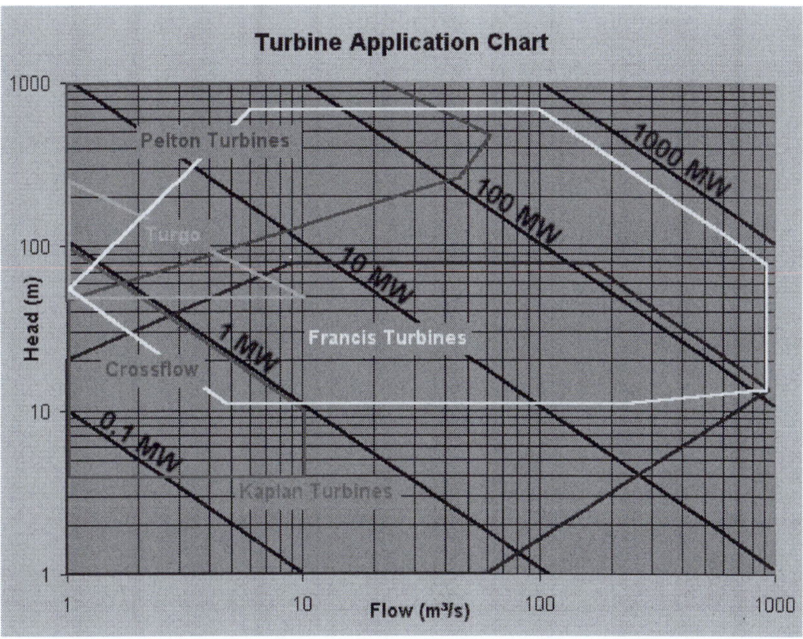

Turbine selection is based on the available water head, and less so on the available flow rate. In general, impulse turbines are used for high head sites, and reaction turbines are used for low head sites. Kaplan turbines with adjustable blade pitch are well-adapted to wide ranges of flow or head conditions, since their peak efficiency can be achieved over a wide range of flow conditions.

Small turbines (mostly under 10 MW) may have horizontal shafts, and even fairly large bulb-type turbines up to 100 MW or so may be horizontal. Very large Francis and Kaplan machines usually have vertical shafts because this makes best use of the available head, and makes installation of a generator more economical. Pelton wheels may be either vertical or horizontal shaft machines because the size of the machine is so much less than the available head. Some impulse turbines use multiple water jets per runner to increase specific speed and balance shaft thrust.

Typical Range of Heads

• Water wheel	$0.2 < H < 4$ (H = head in m)
• Screw turbine	$1 < H < 10$
• VLH turbine	$1.5 < H < 4.5$
• Kaplan turbine	$20 < H < 40$
• Francis turbine	$40 < H < 600$[6]
• Pelton wheel	$50 < H < 1300$
• Turgo turbine	$50 < H < 250$

Specific Speed

The specific speed n_s of a turbine characterizes the turbine's shape in a way that is not related to its size. This allows a new turbine design to be scaled from an existing design of known performance. The specific speed is also the main criteria for matching a specific hydro site with the correct turbine type. The specific speed is the speed with which the turbine turns for a particular discharge Q, with unit head and thereby is able to produce unit power.

Affinity Laws

Affinity laws allow the output of a turbine to be predicted based on model tests. A miniature replica of a proposed design, about one foot (0.3 m) in diameter, can be tested and the laboratory measurements applied to the final application with high confidence. Affinity laws are derived by requiring similitude between the test model and the application.

Flow through the turbine is controlled either by a large valve or by wicket gates arranged around the outside of the turbine runner. Differential head and flow can be plotted for a number of different values of gate opening, producing a hill diagram used to show the efficiency of the turbine at varying conditions.

Runaway Speed

The runaway speed of a water turbine is its speed at full flow, and no shaft load. The turbine will be designed to survive the mechanical forces of this speed. The manufacturer will supply the runaway speed rating.

Maintenance

Turbines are designed to run for decades with very little maintenance of the main elements; overhaul intervals are on the order of several years. Maintenance of the runners and parts exposed to water include removal, inspection, and repair of worn parts.

Normal wear and tear includes pitting corrosion from cavitation, fatigue cracking, and abrasion from suspended solids in the water. Steel elements are repaired by welding, usually with stainless steel rods. Damaged areas are cut or ground out, then welded back up to their original or an improved profile. Old turbine runners may have a significant amount of stainless steel added this way by the end of their lifetime. Elaborate welding procedures may be used to achieve the highest quality repairs.

Other elements requiring inspection and repair during overhauls include bearings, packing box and shaft sleeves, servomotors, cooling systems for the bearings and generator coils, seal rings, wicket gate linkage elements and all surfaces.

Environmental Impact

Water turbines are generally considered a clean power producer, as the turbine causes essentially no change to the water. They use a renewable energy source and are designed to operate for decades. They produce significant amounts of the world's electrical supply.

Historically there have also been negative consequences, mostly associated with the dams normally required for power production. Dams alter the natural ecology of rivers, potentially killing fish, stopping migrations, and disrupting peoples' livelihoods. For example, American Indian tribes in the Pacific Northwest had livelihoods built around salmon fishing, but aggressive dam-building destroyed their way of life. Dams also cause less obvious, but potentially serious consequences, including increased evaporation of water (especially in arid regions), buildup of silt behind the dam, and changes to water temperature and flow patterns. In the United States, it is now illegal to block the migration of fish, for example the endangered white sturgeon in North America, so fish ladders must be provided by dam builders.

Tidal Power

Tidal power, also called tidal energy, is a form of hydropower that converts the energy of tides into useful forms of power, mainly electricity.

Although not yet widely used, tidal power has potential for future electricity generation. Tides are more predictable than wind energy and solar power. Among sources of renewable energy, tidal power has traditionally suffered from relatively high cost and limited availability of sites with sufficiently high tidal ranges or flow velocities, thus constricting its total availability. However, many recent technological developments and improvements, both in design (e.g. dynamic tidal power, tidal lagoons) and turbine technology (e.g. new axial turbines, cross flow turbines), indicate that the total availability of tidal power may be much higher than previously assumed, and that economic and environmental costs may be brought down to competitive levels.

Historically, tide mills have been used both in Europe and on the Atlantic coast of North America. The incoming water was contained in large storage ponds, and as the tide went out, it turned waterwheels that used the mechanical power it produced to mill grain. The earliest occurrences date from the Middle Ages, or even from Roman times. It was only in the 19th century that the process of using falling water and spinning turbines to create electricity was introduced in the U.S. and Europe.

The world's first large-scale tidal power plant is the Rance Tidal Power Station in France, which became operational in 1966.

Generation of Tidal Energy

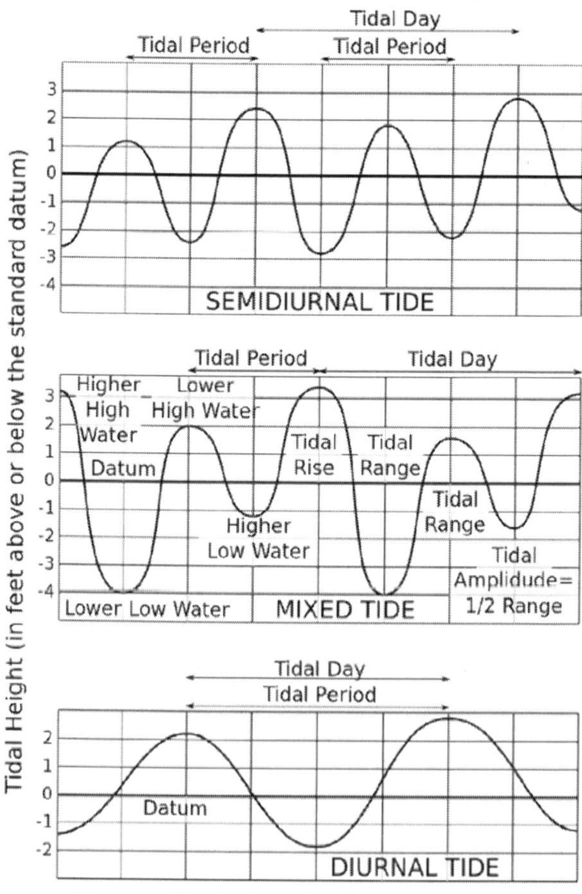

Figure: *Variation of tides over a day*

Tidal power is taken from the Earth's oceanic tides; tidal forces are periodic variations in gravitational attraction exerted by celestial bodies. These forces create corresponding motions or currents in the world's oceans. Due to the strong attraction to the oceans, a bulge in the water level is created, causing a temporary increase in sea level. When the sea level is raised, water from the middle of the ocean is forced to move toward the shorelines, creating a tide.

This occurrence takes place in an unfailing manner, due to the consistent pattern of the moon's orbit around the earth. The magnitude and character of this motion reflects the changing positions of the Moon and Sun relative to the Earth, the effects of Earth's rotation, and local geography of the sea floor and coastlines.

Tidal power is the only technology that draws on energy inherent in the orbital characteristics of the Earth–Moon system, and to a lesser extent in the Earth–Sun system. Other natural energies exploited by human technology originate directly or indirectly with the Sun, including fossil fuel, conventional hydroelectric, wind, biofuel, wave and solar energy. Nuclear energy makes use of Earth's mineral deposits of fissionable elements, while geothermal power taps the Earth's internal heat, which comes from a combination of residual heat from planetary accretion (about 20%) and heat produced through radioactive decay (80%).

A tidal generator converts the energy of tidal flows into electricity. Greater tidal variation and higher tidal current velocities can dramatically increase the potential of a site for tidal electricity generation.

Because the Earth's tides are ultimately due to gravitational interaction with the Moon and Sun and the Earth's rotation, tidal power is practically inexhaustible and classified as a renewable energy resource. Movement of tides causes a loss of mechanical energy in the Earth–Moon system: this is a result of pumping of water through natural restrictions around coastlines and consequent viscous dissipation at the seabed and in turbulence.

This loss of energy has caused the rotation of the Earth to slow in the 4.5 billion years since its formation. During the last 620 million years the period of rotation of the earth (length of a day) has increased from 21.9 hours to 24 hours; in this period the Earth has lost 17% of its rotational energy. While tidal power will take additional energy from the system, the effect is negligible and would only be noticed over millions of years.

Generating Methods

Tidal power can be classified into four generating methods:

Tidal Stream Generator

Tidal stream generators (or TSGs) make use of the kinetic energy of moving water to power turbines, in a similar way to wind turbines that use wind to power turbines. Some tidal generators can be built into the structures of existing bridges, involving virtually no aesthetic problems. Land constrictions such as straits or inlets can create high velocities at specific sites, which can be captured with the use of turbines. These turbines can be horizontal, vertical, open, or ducted and are typically placed near the bottom of the water column.

Tidal Barrage

Tidal barrages make use of the potential energy in the difference in height (or hydraulic head) between high and low tides. When using tidal barrages to generate power, the potential energy from a tide is seized through strategic placement of specialised dams. When the sea level rises and the tide begins to come in, the temporary increase in tidal power is channeled into a large basin behind the dam, holding a large amount of potential energy. With the receding tide, this energy is then converted into mechanical energy as the water is released through large turbines that create electrical power through the use of generators. Barrages are essentially dams across the full width of a tidal estuary.

Dynamic Tidal Power

Dynamic tidal power (or DTP) is an untried but promising technology that would exploit an interaction between potential and kinetic energies in tidal flows. It proposes that very long dams (for example: 30–50 km length) be built from coasts straight out into the sea or ocean, without enclosing an area. Tidal phase differences are introduced across the dam, leading to a significant water-level differential in shallow coastal seas – featuring strong coast-parallel oscillating tidal currents such as found in the UK, China, and Korea.

Tidal Lagoon

A newer tidal energy design option is to construct circular retaining walls embedded with turbines that can capture the potential energy of tides. The created reservoirs are similar to those of tidal barrages, except that the location is artificial and does not contain a preexisting ecosystem.

US and Canadian Studies in the Twentieth Century

The first study of large scale tidal power plants was by the US Federal Power Commission in 1924 which if built would have been located in the northern border area of the US state of Maine and the south eastern border area of the Canadian province of New Brunswick, with various dams, powerhouses, and ship locks enclosing the Bay of Fundy and Passamaquoddy Bay. Nothing came of the study and it is unknown whether Canada had been approached about the study by the US Federal Power Commission.

There was also a report on the international commission in April 1961 entitled " Investigation of the International Passamaquoddy Tidal Power Project" produced by both the US and Canadian Federal Governments. According to benefit to costs ratios, the project was beneficial to the US but not to Canada. A highway system along the top of the dams was envisioned as well.

A study was commissioned by the Canadian, Nova Scotian and New Brunswick Governments (Reassessment of Fundy Tidal Power) to determine the potential for tidal barrages at Chignecto Bay and Minas Basin – at the end of the Fundy Bay estuary. There were three sites determined to be financially feasible: Shepody Bay (1550 MW), Cumberline Basin (1085 MW), and Cobequid Bay (3800 MW). These were never built despite their apparent feasibility in 1977.

Tidal Power Development in the UK

The world's first marine energy test facility was established in 2003 to start the development of the wave and tidal energy industry in the UK. Based in Orkney, Scotland, the European Marine Energy Centre (EMEC) has supported the deployment of more wave and tidal energy devices than at any other single site in the world. EMEC provides a variety of test sites in real sea conditions. Its grid connected tidal test site is located at the Fall of Warness, off the island of Eday, in a narrow channel which concentrates the tide as it flows between the Atlantic Ocean and North Sea. This area has a very strong tidal current, which can travel up to 4 m/s (8 knots) in spring tides. Tidal energy developers currently testing at the site include Alstom (formerly Tidal Generation Ltd), ANDRITZ HYDRO Hammerfest, OpenHydro, Scotrenewables Tidal Power, and Voith.

Current and Future Tidal Power Schemes

- The first tidal power station was the Rance tidal power plant built over a period of 6 years from 1960 to 1966 at La Rance, France. It has 240 MW installed capacity.

- 254 MW Sihwa Lake Tidal Power Plant in South Korea is the largest tidal power installation in the world. Construction was completed in 2011.
- The first tidal power site in North America is the Annapolis Royal Generating Station, Annapolis Royal, Nova Scotia, which opened in 1984 on an inlet of the Bay of Fundy. It has 20 MW installed capacity.
- The Jiangxia Tidal Power Station, south of Hangzhou in China has been operational since 1985, with current installed capacity of 3.2 MW. More tidal power is planned near the mouth of the Yalu River.
- The first in-stream tidal current generator in North America (Race Rocks Tidal Power Demonstration Project) was installed at Race Rocks on southern Vancouver Island in September 2006. The next phase in the development of this tidal current generator will be in Nova Scotia (Bay of Fundy).
- A small project was built by the Soviet Union at Kislaya Guba on the Barents Sea. It has 0.4 MW installed capacity. In 2006 it was upgraded with a 1.2MW experimental advanced orthogonal turbine.
- Jindo Uldolmok Tidal Power Plant in South Korea is a tidal stream generation scheme planned to be expanded progressively to 90 MW of capacity by 2013. The first 1 MW was installed in May 2009.
- A 1.2 MW SeaGen system became operational in late 2008 on Strangford Lough in Northern Ireland.
- The contract for an 812 MW tidal barrage near Ganghwa Island (South Korea) north-west of Incheon has been signed by Daewoo. Completion is planned for 2015.
- A 1,320 MW barrage built around islands west of Incheon is proposed by the South Korean government, with projected construction starting in 2017.
- The Scottish Government has approved plans for a 10MW array of tidal stream generators near Islay, Scotland, costing 40 million pounds, and consisting of 10 turbines – enough to power over 5,000 homes. The first turbine is expected to be in operation by 2013.
- The Indian state of Gujarat is planning to host South Asia's first commercial-scale tidal power station. The company Atlantis Resources planned to install a 50MW tidal farm in the Gulf of

Kutch on India's west coast, with construction starting early in 2012.

- Ocean Renewable Power Corporation was the first company to deliver tidal power to the US grid in September, 2012 when its pilot TidGen system was successfully deployed in Cobscook Bay, near Eastport.
- In New York City, 30 tidal turbines will be installed by Verdant Power in the East River by 2015 with a capacity of 1.05MW.
- Construction of a 240 MW tidal power plant in the city of Swansea in the UK, estimated to begin in Spring 2015. Once completed, it will generate over 400GWh of electricity per year, enough to power roughly 121,000 homes. Completion is scheduled for 2017, and the project has a projected 120 year lifespan.
- A turbine project is being installed in Ramsey Sound in 2014.

Tidal Power Issues

Environmental Concerns: Tidal power can have effects on marine life. The turbines can accidentally kill swimming sea life with the rotating blades. Some fish may no longer utilise the area if threatened with a constant rotating or noise-making object. Marine life is a huge factor when placing tidal power energy generators in the water and precautions are made to ensure that as many marine animals as possible will not be affected by it. The Tethys database provides access to scientific literature and general information on the potential environmental effects of tidal energy.

Tidal Turbines

The main environmental concern with tidal energy is associated with blade strike and entanglement of marine organisms as high speed water increases the risk of organisms being pushed near or through these devices. As with all offshore renewable energies, there is also a concern about how the creation of EMF and acoustic outputs may affect marine organisms. It should be noted that because these devices are in the water, the acoustic output can be greater than those created with offshore wind energy. Depending on the frequency and amplitude of sound generated by the tidal energy devices, this acoustic output can have varying effects on marine mammals (particularly those who echolocate to communicate and navigate in the marine environment such as dolphins and whales). Tidal energy removal can also cause environmental concerns such as degrading farfield water quality and disrupting sediment processes. Depending on the size of the project, these effects can range from small traces of sediment build up near the tidal device to severely affecting nearshore ecosystems and processes.

Tidal Barrage

Installing a barrage may change the shoreline within the bay or estuary, affecting a large ecosystem that depends on tidal flats. Inhibiting the flow of water in and out of the bay, there may also be less flushing of the bay or estuary, causing additional turbidity (suspended solids) and less saltwater, which may result in the death of fish that act as a vital food source to birds and mammals. Migrating fish may also be unable to access breeding streams, and may attempt to pass through the turbines. The same acoustic concerns apply to tidal barrages. Decreasing shipping accessibility can become a socio-economic issue, though locks can be added to allow slow passage. However, the barrage may improve the local economy by increasing land access as a bridge. Calmer waters may also allow better recreation in the bay or estuary.

Tidal Lagoon

Environmentally, the main concerns are blade strike on fish attempting to enter the lagoon, acoustic output from turbines, and changes in sedimentation processes. However, all these effects are localized and do not affect the entire estuary or bay.

Corrosion

Salt water causes corrosion in metal parts. It can be difficult to maintain tidal stream generators due to their size and depth in the water. The use of corrosion-resistant materials such as stainless steels, high-nickel alloys, copper-nickel alloys, nickel-copper alloys and titanium can greatly reduce, or eliminate, corrosion damage.

Mechanical fluids, such as lubricants, can leak out, which may be harmful to the marine life nearby. Proper maintenance can minimize the amount of harmful chemicals that may enter the environment.

Fouling

The biological events that happen when placing any structure in an area of high tidal currents and high biological productivity in the ocean will ensure that the structure becomes an ideal substrate for the growth of marine organisms. In the references of the Tidal Current Project at Race Rocks in British Columbia this is documented. Also see this page and Several structural materials and coatings were tested by the Lester Pearson College divers to assist Clean Current in reducing fouling on the turbine and other underwater infrastructure.

Structural Health Monitoring

The high load factors resulting from the fact that water is 800 times denser than air and the predictable and reliable nature of tides compared with the wind makes tidal energy particularly attractive for Electric power generation. Condition monitoring is the key for exploiting it cost-efficiently.

Wave Power

Wave power is the transport of energy by ocean surface waves, and the capture of that energy to do useful work – for example, electricity generation, water desalination, or the pumping of water (into reservoirs). A machine able to exploit wave power is generally known as a wave energy converter (WEC).

Wave power is distinct from the diurnal flux of tidal power and the steady gyre of ocean currents. Wave-power generation is not currently a widely employed commercial technology, although there have been attempts to use it since at least 1890. In 2008, the first experimental wave farm was opened in Portugal, at the Aguçadoura Wave Park. The major competitor of wave power is offshore wind power.

Physical Concepts

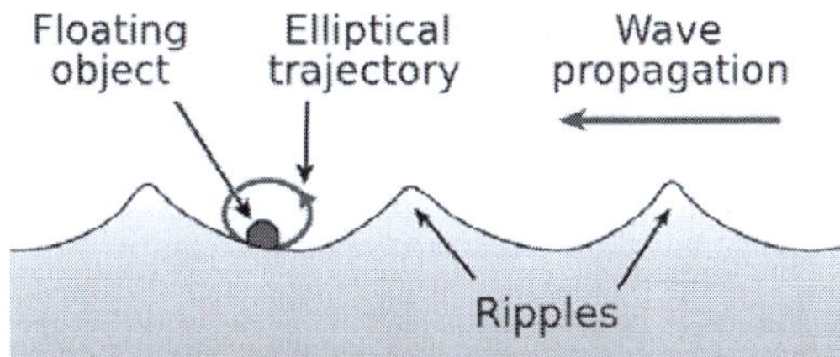

Figure: *When an object bobs up and down on a ripple in a pond, it experiences an elliptical trajectory.*

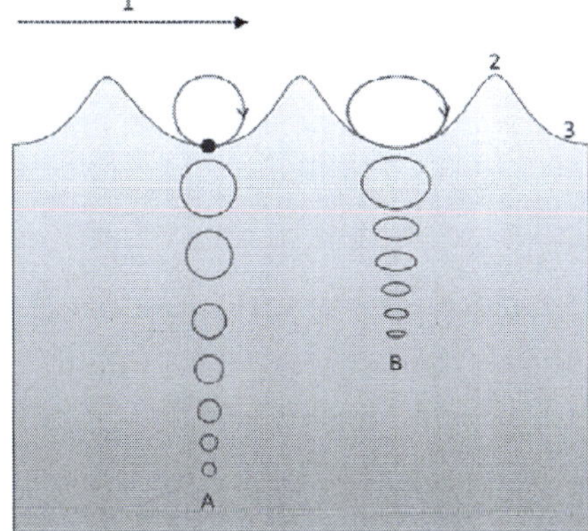

Figure: *Motion of a particle in an ocean wave.*

A = At deep water. The orbital motion of fluid particles decreases rapidly with increasing depth below the surface.

B = At shallow water (ocean floor is now at B). The elliptical movement of a fluid particle flattens with decreasing depth.

1 = Propagation direction.

2 = Wave crest.

3 = Wave trough.

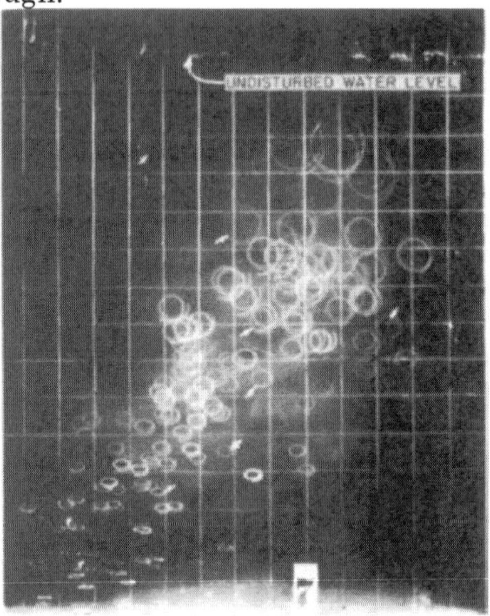

Figure: *Photograph of the water particle orbits under a – progressive and periodic – surface gravity wave in a wave flume. The wave conditions are: mean water depth* d = *2.50 ft (0.76 m), wave height* H = *0.339 ft (0.103 m), wavelength* λ = *6.42 ft (1.96 m), period* T = *1.12 s.*

Waves are generated by wind passing over the surface of the sea. As long as the waves propagate slower than the wind speed just above the waves, there is an energy transfer from the wind to the waves. Both air pressure differences between the upwind and the lee side of a wave crest, as well as friction on the water surface by the wind, making the water to go into the shear stress causes the growth of the waves.

Wave height is determined by wind speed, the duration of time the wind has been blowing, fetch (the distance over which the wind excites the waves) and by the depth and topography of the seafloor (which can focus or disperse the energy of the waves). A given wind speed has a matching practical limit over which time or distance will not produce larger waves. When this limit has been reached the sea is said to be "fully developed".

In general, larger waves are more powerful but wave power is also determined by wave speed, wavelength, and water density.

Oscillatory motion is highest at the surface and diminishes exponentially with depth. However, for standing waves (clapotis) near a reflecting coast, wave energy is also present as pressure oscillations at great depth, producing microseisms. These pressure fluctuations at greater depth are too small to be interesting from the point of view of wave power.

The waves propagate on the ocean surface, and the wave energy is also transported horizontally with the group velocity. The mean transport rate of the wave energy through a vertical plane of unit width, parallel to a wave crest, is called the wave energy flux (or wave power, which must not be confused with the actual power generated by a wave power device).

Wave Power Formula

In deep water where the water depth is larger than half the wavelength, the wave energy flux is

$$P = \frac{\rho g^2}{64\pi} H_{m0}^2 T_e \approx \left(0.5 \frac{\text{kW}}{\text{m}^3 \cdot \text{s}}\right) H_{m0}^2 \, T_e,$$

with P the wave energy flux per unit of wave-crest length, H_{m0} the significant wave height, T_e the wave energy period, ρ the water density and g the acceleration by gravity. The above formula states that wave power is proportional to the wave energy period and to the square of the wave height. When the significant wave height is given in metres, and the wave period in seconds, the result is the wave power in kilowatts (kW) per metre of wavefront length.

Example: Consider moderate ocean swells, in deep water, a few km off a coastline, with a wave height of 3 m and a wave energy period of 8 seconds. Using the formula to solve for power, we get

$$P \approx 0.5 \frac{\text{kW}}{\text{m}^3 \cdot \text{s}} (3 \cdot \text{m})^2 (8 \cdot \text{s}) \approx 36 \frac{\text{kW}}{\text{m}},$$

meaning there are 36 kilowatts of power potential per meter of wave crest.

In major storms, the largest waves offshore are about 15 meters high and have a period of about 15 seconds. According to the above formula, such waves carry about 1.7 MW of power across each metre of wavefront.

An effective wave power device captures as much as possible of the wave energy flux. As a result the waves will be of lower height in the region behind the wave power device.

Wave Energy and Wave-Energy Flux

In a sea state, the average(mean) energy density per unit area of gravity waves on the water surface is proportional to the wave height squared, according to linear wave theory:

$$E = \frac{1}{8}\rho g H_{m0}^2,$$

where E is the mean wave energy density per unit horizontal area (J/m^2), the sum of kinetic and potential energy density per unit horizontal area. The potential energy density is equal to the kinetic energy, both contributing half to the wave energy density E, as can be expected from the equipartition theorem. In ocean waves, surface tension effects are negligible for wavelengths above a few decimetres.

As the waves propagate, their energy is transported. The energy transport velocity is the group velocity. As a result, the wave energy flux, through a vertical plane of unit width perpendicular to the wave propagation direction, is equal to:

$$P = E c_g,$$

with c_g the group velocity (m/s). Due to the dispersion relation for water waves under the action of gravity, the group velocity depends on the wavelength λ, or equivalently, on the wave period T. Further, the dispersion relation is a function of the water depth h. As a result, the group velocity behaves differently in the limits of deep and shallow water, and at intermediate depths:

Deep-Water Characteristics and Opportunities

Deep water corresponds with a water depth larger than half the wavelength, which is the common situation in the sea and ocean. In deep water, longer-period waves propagate faster and transport their energy faster. The deep-water group velocity is half the phase velocity. In shallow water, for wavelengths larger than about twenty times the water depth, as found quite often near the coast, the group velocity is equal to the phase velocity.

History

The first known patent to use energy from ocean waves dates back to 1799, and was filed in Paris by Girard and his son. An early application of wave power was a device constructed around 1910 by

Bochaux-Praceique to light and power his house at Royan, near Bordeaux in France. It appears that this was the first oscillating water-column type of wave-energy device. From 1855 to 1973 there were already 340 patents filed in the UK alone.

Modern scientific pursuit of wave energy was pioneered by Yoshio Masuda's experiments in the 1940s. He has tested various concepts of wave-energy devices at sea, with several hundred units used to power navigation lights. Among these was the concept of extracting power from the angular motion at the joints of an articulated raft, which was proposed in the 1950s by Masuda.

A renewed interest in wave energy was motivated by the oil crisis in 1973. A number of university researchers re-examined the potential to generate energy from ocean waves, among whom notably were Stephen Salter from the University of Edinburgh, Kjell Budal and Johannes Falnes from Norwegian Institute of Technology (now merged into Norwegian University of Science and Technology), Michael E. McCormick from U.S. Naval Academy, David Evans from Bristol University, Michael French from University of Lancaster, Nick Newman and C. C. Mei from MIT.

Stephen Salter's 1974 invention became known as Salter's duck or *nodding duck*, although it was officially referred to as the Edinburgh Duck. In small scale controlled tests, the Duck's curved cam-like body can stop 90% of wave motion and can convert 90% of that to electricity giving 81% efficiency.

In the 1980s, as the oil price went down, wave-energy funding was drastically reduced. Nevertheless, a few first-generation prototypes were tested at sea. More recently, following the issue of climate change, there is again a growing interest worldwide for renewable energy, including wave energy.

The world's first marine energy test facility was established in 2003 to kick start the development of a wave and tidal energy industry in the UK. Based in Orkney, Scotland, the European Marine Energy Centre (EMEC) has supported the deployment of more wave and tidal energy devices than at any other single site in the world. EMEC provides a variety of test sites in real sea conditions. It's grid connected wave test site is situated at Billia Croo, on the western edge of the Orkney mainland, and is subject to the full force of the Atlantic Ocean with seas as high as 19 metres recorded at the site. Wave energy developers currently testing at the centre include Aquamarine Power, Pelamis Wave Power, ScottishPower Renewables and Wello.

Modern Technology

Wave power devices are generally categorized by the method used to capture the energy of the waves, by location and by the power take-off system. Locations are shoreline, nearshore and offshore. Types of power take-off include: hydraulic ram, elastomeric hose pump, pump-to-shore, hydroelectric turbine, air turbine, and linear electrical generator. When evaluating wave energy as a technology type, it is important to distinguish between the four most common approaches: point absorber buoys, surface attenuators, oscillating water columns, and overtopping devices.

Point Absorber Buoy

This device floats on the surface of the water, held in place by cables connected to the seabed. Buoys use the rise and fall of swells to drive hydraulic pumps and generate electricity. EMF generated by electrical transmission cables and acoustic of these devices may be a concern for marine organisms. The presence of the buoys may affect fish, marine mammals, and birds as potential minor collision risk and roosting sites. Potential also exists for entanglement in mooring lines. Energy removed from the waves may also affect the shoreline, resulting in a recommendation that sites remain a considerable distance from the shore.

Surface Attenuator

These devices act similarly to point absorber buoys, with multiple floating segments connected to one another and are oriented perpendicular to incoming waves. A flexing motion is created by swells that drive hydraulic pumps to generate electricity. Environmental effects are similar to those of point absorber buoys, with an additional concern that organisms could be pinched in the joints.

Oscillating Water Column

Oscillating water column devices can be located on shore or in deeper waters offshore. With an air chamber integrated into the device, swells compress air in the chambers forcing air through an air turbine to create electricity. Significant noise is produced as air is pushed through the turbines, potentially affecting birds and other marine organisms within the vicinity of the device. There is also concern about marine organisms getting trapped or entangled within the air chambers.

Overtopping Device

Overtopping devices are long structures that use wave velocity to fill a reservoir to a greater water level than the surrounding ocean.

The potential energy in the reservoir height is then captured with low-head turbines. Devices can be either on shore or floating offshore. Floating devices will have environmental concerns about the mooring system affecting benthic organisms, organisms becoming entangled, or EMF effects produced from subsea cables. There is also some concern regarding low levels of turbine noise and wave energy removal affecting the nearfield habitat.

Oscillating Wave Surge Converter

These devices typically have one end fixed to a structure or the seabed while the other end is free to move. Energy is collected from the relative motion of the body compared to the fixed point. Oscillating wave surge converters often come in the form of floats, flaps, or membranes. Environmental concerns include minor risk of collision, artificial reefing near the fixed point, EMF effects from subsea cables, and energy removal effecting sediment transport. Some of these designs incorporate parabolic reflectors as a means of increasing the wave energy at the point of capture. These capture systems use the rise and fall motion of waves to capture energy. Once the wave energy is captured at a wave source, power must be carried to the point of use or to a connection to the electrical grid by transmission power cables.

Potential

The worldwide resource of wave energy has been estimated to be greater than 2 TW. Locations with the most potential for wave power include the western seaboard of Europe, the northern coast of the UK, and the Pacific coastlines of North and South America, Southern Africa, Australia, and New Zealand. The north and south temperate zones have the best sites for capturing wave power. The prevailing westerlies in these zones blow strongest in winter.

Challenges

There is a potential impact on the marine environment. Noise pollution, for example, could have negative impact if not monitored, although the noise and visible impact of each design vary greatly. Other biophysical impacts (flora and fauna, sediment regimes and water column structure and flows) of scaling up the technology is being studied. In terms of socio-economic challenges, wave farms can result in the displacement of commercial and recreational fishermen from productive fishing grounds, can change the pattern of beach sand nourishment, and may represent hazards to safe navigation. Waves generate about 2,700 gigawatts of power. Of those 2,700 gigawatts, only about 500 gigawatts can be captured with the current technology.

Wave Farms

Portugal:

- The Aguçadoura Wave Farm was the world's first wave farm. It was located 5 km (3 mi) offshore near Póvoa de Varzim, north of Porto, Portugal. The farm was designed to use three Pelamis wave energy converters to convert the motion of the ocean surface waves into electricity, totalling to 2.25 MW in total installed capacity. The farm first generated electricity in July 2008 and was officially opened on September 23, 2008, by the Portuguese Minister of Economy. The wave farm was shut down two months after the official opening in November 2008 as a result of the financial collapse of Babcock & Brown due to the global economic crisis. The machines were off-site at this time due to technical problems, and although resolved have not returned to site and were subsequently scrapped in 2011 as the technology had moved on to the P2 variant as supplied to Eon and Scottish Power Renewables. A second phase of the project planned to increase the installed capacity to 21 MW using a further 25 Pelamis machines is in doubt following Babcock's financial collapse.

United Kingdom:

- Funding for a 3 MW wave farm in Scotland was announced on February 20, 2007, by the Scottish Executive, at a cost of over 4 million pounds, as part of a £13 million funding package for marine power in Scotland. The first of 66 machines was launched in May 2010.
- A facility known as Wave hub has been constructed off the north coast of Cornwall, England, to facilitate wave energy development. The Wave hub will act as giant extension cable, allowing arrays of wave energy generating devices to be connected to the electricity grid. The Wave hub will initially allow 20 MW of capacity to be connected, with potential expansion to 40 MW. Four device manufacturers have so far expressed interest in connecting to the Wave hub. The scientists have calculated that wave energy gathered at Wave Hub will be enough to power up to 7,500 households. The site has the potential to save greenhouse gas emissions of about 300,000 tons of carbon dioxide in the next 25 years.

Australia:

- A CETO wave farm off the coast of Western Australia has been operating to prove commercial viability and, after preliminary environmental approval, is poised for further development.

- Ocean Power Technologies (OPT Australasia Pty Ltd) is developing a wave farm connected to the grid near Portland, Victoria through a 19 MW wave power station. The project has received an AU $66.46 million grant from the Federal Government of Australia.
- Oceanlinx will deploy a commercial scale demonstrator off the coast of South Australia at Port MacDonnell before the end of 2013. This device, the *greenWAVE*, has a rated electrical capacity of 1MW. This project has been supported by ARENA through the Emerging Renewables Program. The *greenWAVE* device is a bottom standing gravity structure, that does not require anchoring or seabed preparation and with no moving parts below the surface of the water.

United States:

- Reedsport, Oregon – a commercial wave park on the west coast of the United States located 2.5 miles offshore near Reedsport, Oregon. The first phase of this project is for ten PB150 PowerBuoys, or 1.5 megawatts. The Reedsport wave farm was scheduled for installation spring 2013. In 2013, the project has ground to a halt because of legal and technical problems.

Wind Power

Wind power or wind energy is the energy extracted from wind using wind turbines to produce electrical power, windmills for mechanical power, windpumps for water pumping, or sails to propel ships.

Large wind farms consist of hundreds of individual wind turbines which are connected to the electric power transmission network. According to the recent EU analysis for new constructions, onshore wind is an inexpensive source of electricity, competitive with or in many places cheaper than coal, gas or fossil fuel plants. Offshore wind is steadier and stronger than on land, and offshore farms have less visual impact, but construction and maintenance costs are considerably higher. Small onshore wind farms can feed some energy into the grid or provide electricity to isolated off-grid locations.

Wind power, as an alternative to fossil fuels, is plentiful, renewable, widely distributed, clean, produces no greenhouse gas emissions during operation and uses little land. The effects on the environment are generally less problematic than those from other power sources. As of 2013, Denmark is generating more than a third of its electricity from wind and 83 countries around the world are using wind power to supply the electricity grid. Wind power capacity has expanded rapidly to 336

GW in June 2014, and wind energy production was around 4% of total worldwide electricity usage, and growing rapidly.

Wind power is very consistent from year to year but has significant variation over shorter time scales. As the proportion of windpower in a region increases, a need to upgrade the grid, and a lowered ability to supplant conventional production can occur. Power management techniques such as having excess capacity storage, geographically distributed turbines, dispatchable backing sources, storage such as pumped-storage hydroelectricity, exporting and importing power to neighbouring areas or reducing demand when wind production is low, can greatly mitigate these problems. In addition, weather forecasting permits the electricity network to be readied for the predictable variations in production that occur. Wind power can be considered a topic in *applied eolics*.

History

Wind power has been used as long as humans have put sails into the wind. For more than two millennia wind-powered machines have ground grain and pumped water. Wind power was widely available and not confined to the banks of fast-flowing streams, or later, requiring sources of fuel. Wind-powered pumps drained the polders of the Netherlands, and in arid regions such as the American mid-west or the Australian outback, wind pumps provided water for live stock and steam engines.

The first windmill used for the production of electricity was built in Scotland in July 1887 by Prof James Blyth of Anderson's College, Glasgow (the precursor of Strathclyde University). Blyth's 10 m high, cloth-sailed wind turbine was installed in the garden of his holiday cottage at Marykirk in Kincardineshire and was used to charge accumulators developed by the Frenchman Camille Alphonse Faure, to power the lighting in the cottage, thus making it the first house in the world to have its electricity supplied by wind power. Blyth offered the surplus electricity to the people of Marykirk for lighting the main street, however, they turned down the offer as they thought electricity was "the work of the devil." Although he later built a wind turbine to supply emergency power to the local Lunatic Asylum, Infirmary and Dispensary of Montrose the invention never really caught on as the technology was not considered to be economically viable.

Across the Atlantic, in Cleveland, Ohio a larger and heavily engineered machine was designed and constructed in the winter of 1887-1888 by Charles F. Brush, this was built by his engineering company at his home and operated from 1886 until 1900. The Brush wind turbine had a rotor 17 m (56 foot) in diameter and was mounted on an 18 m (60 foot) tower. Although large by today's standards, the machine was only rated at 12 kW. The connected dynamo was used either to charge a bank of batteries or to operate up to 100 incandescent light bulbs, three arc lamps, and various motors in Brush's laboratory.

With the development of electric power, wind power found new applications in lighting buildings remote from centrally-generated power. Throughout the 20th century parallel paths developed small wind stations suitable for farms or residences, and larger utility-scale wind generators that could be connected to electricity grids for remote use of power. Today wind powered generators operate in every size range between tiny stations for battery charging at isolated residences, up to near-gigawatt sized offshore wind farms that provide electricity to national electrical networks.

Wind Farms

Large onshore wind farms		
Wind farm	***Current capacity (MW)***	***Country***
Gansu Wind Farm	6,000	China
Alta (Oak Creek-Mojave)	1,320	United States
Jaisalmer Wind Park	1,064	India
Shepherds Flat Wind Farm	845	United States
Roscoe Wind Farm	782	United States
Horse Hollow Wind Energy Center	736	United States
Capricorn Ridge Wind Farm	662	United States
Fântânele-Cogealac Wind Farm	600	Romania
Fowler Ridge Wind Farm	600	United States
Whitelee Wind Farm	539	United Kingdom

A wind farm is a group of wind turbines in the same location used for production of electricity. A large wind farm may consist of several hundred individual wind turbines distributed over an extended area, but the land between the turbines may be used for agricultural or other purposes. A wind farm may also be located offshore.

Almost all large wind turbines have the same design — a horizontal axis wind turbine having an upwind rotor with three blades, attached to a nacelle on top of a tall tubular tower.

In a wind farm, individual turbines are interconnected with a medium voltage (often 34.5 kV), power collection system and communications network. At a substation, this medium-voltage electric current is increased in voltage with a transformer for connection to the high voltage electric power transmission system.

Offshore Wind Power

Offshore wind power refers to the construction of wind farms in large bodies of water to generate electricity. These installations can utilise the more frequent and powerful winds that are available in these locations and have less aesthetic impact on the landscape than land based projects. However, the construction and the maintenance costs are considerably higher.

Siemens and Vestas are the leading turbine suppliers for offshore wind power. DONG Energy, Vattenfall and E.ON are the leading offshore operators. As of October 2010, 3.16 GW of offshore wind power capacity was operational, mainly in Northern Europe. According to BTM Consult, more than 16 GW of additional capacity will be installed before the end of 2014 and the UK and Germany will become the two leading markets. Offshore wind power capacity is expected to reach a total of 75 GW worldwide by 2020, with significant contributions from China and the US.

At the end of 2012, 1,662 turbines at 55 offshore wind farms in 10 European countries are generating 18 TWh, which can power almost five million households. As of August 2013 the London Array in the United Kingdom is the largest offshore wind farm in the world at 630 MW. This is followed by the Greater Gabbard Wind Farm (504 MW), also in the UK. The Gwynt y Mτr wind farm (576 MW) is the largest project currently under construction.

Wind Power Capacity and Production

Worldwide there are now over two hundred thousand wind turbines operating, with a total nameplate capacity of 282,482 MW as of end 2012. The European Union alone passed some 100,000 MW nameplate capacity in September 2012, while the United States surpassed 50,000 MW in August 2012 and China's grid connected capacity passed 50,000 MW the same month.

World wind generation capacity more than quadrupled between 2000 and 2006, doubling about every three years. The United States pioneered wind farms and led the world in installed capacity in the 1980s and into the 1990s. In 1997 German installed capacity surpassed the U.S. and led until once again overtaken by the U.S. in 2008. China has been rapidly expanding its wind installations in the late 2000s and passed the U.S. in 2010 to become the world leader.

Wind power capacity has expanded rapidly to 336 GW in June 2014, and wind energy production was around 4% of total worldwide electricity usage, and growing rapidly.

The actual amount of electricity that wind is able to generate is calculated by multiplying the nameplate capacity by the capacity factor, which varies according to equipment and location. Estimates of the capacity factors for wind installations are in the range of 35% to 44%.

Several countries have already achieved relatively high levels of penetration, such as 33.2% of stationary (grid) electricity production in Denmark (2013), 19% in Portugal (2011), and 16% in Spain (2011).

In Australia, the state of South Australia generates around half of the nation's wind power capacity. By the end of 2011 wind power in South Australia, championed by Premier (and Climate Change Minister) Mike Rann, reached 26% of the State's electricity generation, edging out coal for the first time. At this stage South Australia, with only 7.2% of Australia's population, had 54% of Australia's installed capacity. 16% in Ireland (2012) and 8% in Germany (2011). As of 2011, 83 countries around the world were using wind power on a commercial basis.

Europe accounted for 48% of the world total wind power generation capacity in 2009. In 2010, Spain became Europe's leading producer of wind energy, achieving 42,976 GWh. Germany held the top spot in Europe in terms of installed capacity, with a total of 27,215 MW as of 31 December 2010.

Top 10 countries by nameplate windpower capacity (2013 year-end)			
Country	***New 2013 capacity (MW)***	***Windpower total capacity (MW)***	***% world total***
China	16,088	91,412	28.7
United States	1,084	61,091	19.2
Germany	3,238	34,250	10.8
Spain	175	22,959	7.2
India	1,729	20,150	6.3
UK	1,883	10,531	3.3
Italy	444	8,552	2.7
France	631	8,254	2.6
Canada	1,599	7,803	2.5
Denmark	657	4,772	1.5
(rest of world)	7,761	48,332	15.2
World total	35,289 MW	318,105 MW	100%

Top 10 countries by windpower electricity production (2012 totals)		
Country	***Windpower production (TWh)***	***% world total***
United States	140.9	26.4
China	118.1	22.1
Spain	49.1	9.2
Germany	46.0	8.6
India	30.0	5.6
UK	19.6	3.7
France	14.9	2.8
Italy	13.4	2.5
Canada	11.8	2.2
Denmark	10.3	1.9
(rest of world)	80.2	15.0
World total	534.3 TWh	100%

Growth Trends

In 2010, more than half of all new wind power was added outside of the traditional markets in Europe and North America. This was largely from new construction in China, which accounted for nearly half the new wind installations (16.5 GW).

Global Wind Energy Council (GWEC) figures show that 2007 recorded an increase of installed capacity of 20 GW, taking the total installed wind energy capacity to 94 GW, up from 74 GW in 2006. Despite constraints facing supply chains for wind turbines, the annual market for wind continued to increase at an estimated rate of 37%, following 32% growth in 2006. In terms of economic value, the wind energy sector has become one of the important players in the energy markets, with the total value of new generating equipment installed in 2007 reaching €25 billion, or US$36 billion.

Although the wind power industry was affected by the global financial crisis in 2009 and 2010, a BTM Consult five-year forecast up to 2013 projects substantial growth. Over the past five years the average growth in new installations has been 27.6% each year. In the forecast to 2013 the expected average annual growth rate is 15.7%. More than 200 GW of new wind power capacity could come on line before the end of 2014. Wind power market penetration is expected to reach 3.35% by 2013 and 8% by 2018.

Capacity Factor

Since wind speed is not constant, a wind farm's annual energy production is never as much as the sum of the generator nameplate ratings multiplied by the total hours in a year. The ratio of actual productivity in a year to this theoretical maximum is called the capacity factor. Typical capacity factors are 15–50%; values at the upper end of the range are achieved in favourable sites and are due to wind turbine design improvements. Online data is available for some locations, and the capacity factor can be calculated from the yearly output. For example, the German nation-wide average wind power capacity factor over all of 2012 was just under 17.5% (45867 GW·h/yr / (29.9 GW × 24 × 366) = 0.1746), and the capacity factor for Scottish wind farms averaged 24% between 2008 and 2010.

Unlike fuelled generating plants, the capacity factor is affected by several parameters, including the variability of the wind at the site and the size of the generator relative to the turbine's swept area. A small generator would be cheaper and achieve a higher capacity factor but would produce less electricity (and thus less profit) in high winds. Conversely, a large generator would cost more but generate little extra power and, depending on the type, may stall out at low wind speed. Thus an optimum capacity factor of around 40–50% would be aimed for.

A 2008 study released by the U.S. Department of Energy noted that the capacity factor of new wind installations was increasing as the technology improves, and projected further improvements for future capacity factors. In 2010, the department estimated the capacity factor of new wind turbines in 2010 to be 45%. The overall average capacity factor for wind generation in the US increased from 31.7% in 2008, to 32.3% in 2013.

Penetration

Wind energy penetration refers to the fraction of energy produced by wind compared with the total available generation capacity. There is no generally accepted maximum level of wind penetration. The limit for a particular grid will depend on the existing generating plants, pricing mechanisms, capacity for energy storage, demand management and other factors. An interconnected electricity grid will already include reserve generating and transmission capacity to allow for equipment failures. This reserve capacity can also serve to compensate for the varying power generation produced by wind stations. Studies have indicated that 20% of the total annual electrical energy consumption may be incorporated with minimal difficulty. These studies have been for locations with geographically dispersed wind farms, some degree of dispatchable energy or hydropower with storage capacity, demand management, and interconnected to a large grid area enabling the export of electricity when needed. Beyond the 20% level, there are few technical limits, but the economic implications become more significant.

Electrical utilities continue to study the effects of large scale penetration of wind generation on system stability and economics.

A wind energy penetration figure can be specified for different durations of time. On an annual basis, as of 2011, few grid systems have penetration levels above 5%: Denmark – 29%, Portugal – 19%, Spain – 19%, Ireland – 18%, and Germany – 11%. For the U.S. in 2011, the penetration level was estimated at 3.3%. To obtain 100% from wind annually requires substantial long term storage. On a monthly, weekly, daily, or hourly basis—or less—wind can supply as much as or more than 100% of current use, with the rest stored or exported. Seasonal industry can take advantage of high wind and low usage times such as at night when wind output can exceed normal demand. Such industry can include production of silicon, aluminum, steel, or of natural gas, and hydrogen, which allow long term storage, facilitating 100% energy from variable renewable energy. Homes can also be programmed to accept extra electricity on demand, for example by remotely turning up water heater thermostats.

Variability

Electricity generated from wind power can be highly variable at several different timescales: hourly, daily, or seasonally. Annual variation also exists, but is not as significant. Because instantaneous electrical generation and consumption must remain in balance to maintain grid stability, this variability can present substantial challenges to incorporating large amounts of wind power into a grid system. Intermittency and the non-dispatchable nature of wind energy production can raise costs for regulation, incremental operating reserve, and (at high penetration levels) could require an increase in the already existing energy demand management, load shedding, storage solutions or system interconnection with HVDC cables.

Fluctuations in load and allowance for failure of large fossil-fuel generating units require reserve capacity that can also compensate for variability of wind generation.

Increase in system operation costs, Euros per MWh, for 10% & 20% wind share

Increase in system operation costs, Euros per MWh, for 10% & 20% wind share		
Country	***10%***	***20%***
Germany	2.5	3.2
Denmark	0.4	0.8
Finland	0.3	1.5
Norway	0.1	0.3
Sweden	0.3	0.7

Wind power is however, variable, but during low wind periods it can be replaced by other power sources. Transmission networks presently cope with outages of other generation plants and daily changes in electrical demand, but the variability of intermittent power sources such as wind power, are unlike those of conventional power generation plants which, when scheduled to be operating, may be able to deliver their nameplate capacity around 95% of the time.

Presently, grid systems with large wind penetration require a small increase in the frequency of usage of natural gas spinning reserve power plants to prevent a loss of electricity in the event that conditions are not favourable for power production from the wind. At lower wind power grid penetration, this is less of an issue.

GE has installed a prototype wind turbine with onboard battery similar to that of an electric car, equivalent of 1 minute of production. Despite the small capacity, it is enough to guarantee that power output complies with forecast for 15 minutes, as the battery is used to eliminate the difference rather than provide full output. The increased predictability can be used to take wind power penetration from 20 to 30 or 40 per cent. The battery cost can be retrieved by selling burst power on demand and reducing backup needs from gas plants.

A report on Denmark's wind power noted that their wind power network provided less than 1% of average demand on 54 days during the year 2002. Wind power advocates argue that these periods of low wind can be dealt with by simply restarting existing power stations that have been held in readiness, or interlinking with HVDC. Electrical grids with slow-responding thermal power plants and without ties to networks with hydroelectric generation may have to limit the use of wind power. According to a 2007 Stanford University study published in the Journal of Applied Meteorology and Climatology, interconnecting ten or more wind farms can allow an average of 33% of the total energy produced (i.e. about 8% of total nameplate capacity) to be used as reliable, baseload electric power which can be relied on to handle peak loads, as long as minimum criteria are met for wind speed and turbine height.

Conversely, on particularly windy days, even with penetration levels of 16%, wind power generation can surpass all other electricity sources in a country. In Spain, on 16 April 2012 wind power production reached the highest percentage of electricity production till then, with wind farms covering 60.46% of the total demand. In Denmark, which had power market penetration of 30% in 2013, over 90 hours, wind power generated 100% of the countries power, peaking at 122% of the countries demand at 2 am on 28 October.

A 2006 International Energy Agency forum presented costs for managing intermittency as a function of wind-energy's share of total capacity for several countries, as shown in the table on the right. Three reports on the wind variability in the UK issued in 2009, generally agree that variability of wind needs to be taken into account, but it does not make the grid unmanageable. The additional costs, which are modest, can be quantified.

The combination of diversifying variable renewables by type and location, forecasting their variation, and integrating them with dispatchable renewables, flexible fuelled generators, and demand response can create a power system that has the potential to meet power supply needs reliably. Integrating ever-higher levels of renewables is being successfully demonstrated in the real world:

> *In 2009, eight American and three European authorities, writing in the leading electrical engineers' professional journal, didn't find "a credible and firm technical limit to the amount of wind energy that can be accommodated by electricity grids". In fact, not one of more than 200 international studies, nor official studies for the eastern and western U.S. regions, nor the International Energy Agency, has found major costs or technical barriers to reliably integrating up to 30% variable renewable supplies into the grid, and in some studies much more.*
>
> – *Reinventing Fire*

Solar power tends to be complementary to wind. On daily to weekly timescales, high pressure areas tend to bring clear skies and low surface winds, whereas low pressure areas tend to be windier and cloudier. On seasonal timescales, solar energy peaks in summer, whereas in many areas wind energy is lower in summer and higher in winter. Thus the intermittencies of wind and solar power tend to cancel each other somewhat. In 2007 the Institute for Solar Energy Supply Technology of the University of Kassel pilot-tested a combined power plant linking solar, wind, biogas and hydrostorage to provide load-following power around the clock and throughout the year, entirely from renewable sources.

Predictability

Wind power forecasting methods are used, but predictability of any particular wind farm is low for short-term operation. For any particular generator there is an 80% chance that wind output will change less than 10% in an hour and a 40% chance that it will change 10% or more in 5 hours.

However, studies by Graham Sinden (2009) suggest that, in practice, the variations in thousands of wind turbines, spread out over several different sites and wind regimes, are smoothed. As the distance between sites increases, the correlation between wind speeds measured at those sites, decreases.

Thus, while the output from a single turbine can vary greatly and rapidly as local wind speeds vary, as more turbines are connected over larger and larger areas the average power output becomes less variable and more predictable.

Wind power hardly ever suffers major technical failures, since failures of individual wind turbines have hardly any effect on overall power, so that the distributed wind power is reliable and predictable, whereas conventional generators, while far less variable, can suffer major unpredictable outages.

Energy Storage

Typically, conventional hydroelectricity complements wind power very well. When the wind is blowing strongly, nearby hydroelectric stations can temporarily hold back their water. When the wind drops they can, provided they have the generation capacity, rapidly increase production to compensate. This gives a very even overall power supply and virtually no loss of energy and uses no more water.

Alternatively, where a suitable head of water is not available, pumped-storage hydroelectricity or other forms of grid energy storage such as compressed air energy storage and thermal energy storage can store energy developed by high-wind periods and release it when needed. The type of storage needed depends on the wind penetration level – low penetration requires daily storage, and high penetration requires both short and long term storage – as long as a month or more. Stored energy increases the economic value of wind energy since it can be shifted to displace higher cost generation during peak demand periods. The potential revenue from this arbitrage can offset the cost and losses of storage; the cost of storage may add 25% to the cost of any wind energy stored but it is not envisaged that this would apply to a large proportion of wind energy generated. For example, in the UK, the 1.7 GW Dinorwig pumped-storage plant evens out electrical demand peaks, and allows base-load suppliers to run their plants more efficiently. Although pumped-storage power systems are only about 75% efficient, and have high installation costs, their low running costs and ability to reduce the required electrical base-load can save both fuel and total electrical generation costs.

In particular geographic regions, peak wind speeds may not coincide with peak demand for electrical power. In the U.S. states of California and Texas, for example, hot days in summer may have low wind speed and high electrical demand due to the use of air conditioning. Some utilities subsidize the purchase of geothermal heat pumps by their customers, to reduce electricity demand during the summer months by making air conditioning up to 70% more efficient; widespread adoption of this technology would better match electricity demand to wind availability in areas with hot summers and low summer winds. A possible future option may be to interconnect widely dispersed geographic areas with an HVDC "super grid". In the U.S. it is estimated that to upgrade the transmission system to take in planned or potential renewables would cost at least $60 billion.

Germany has an installed capacity of wind and solar that can exceed daily demand, and has been exporting peak power to neighbouring countries, with exports which amounted to some 14.7 billion kilowatt hours in 2012. A more practical solution is the installation of thirty days storage capacity able to supply 80% of demand, which will become necessary when most of Europe's energy is obtained from wind power and solar power. Just as the EU requires member countries to maintain 90 days strategic reserves of oil it can be expected that countries will provide electricity storage, instead of expecting to use their neighbours for net metering.

Capacity Credit and Fuel Savings

The capacity credit of wind is estimated by determining the capacity of conventional plants displaced by wind power, whilst maintaining the same degree of system security,. However, the precise value is irrelevant since the main value of wind is its fuel and CO_2 savings, and wind is not expected to be constantly available.

Economics

Wind turbines reached grid parity (the point at which the cost of wind power matches traditional sources) in some areas of Europe in the mid-2000s, and in the US around the same time. Falling prices continue to drive the levelized cost down and it has been suggested that it has reached general grid parity in Europe in 2010, and will reach the same point in the US around 2016 due to an expected reduction in capital costs of about 12%.

Electricity Cost and Trends

Wind power is capital intensive, but has no fuel costs. The price of wind power is therefore much more stable than the volatile prices of fossil fuel sources. The marginal cost of wind energy once a station is constructed is usually less than 1-cent per kW ·h.

However, the estimated average cost per unit of electricity must incorporate the cost of construction of the turbine and transmission facilities, borrowed funds, return to investors (including cost of risk), estimated annual production, and other components, averaged over the projected useful life of the equipment, which may be in excess of twenty years. Energy cost estimates are highly dependent on these assumptions so published cost figures can differ substantially. In 2004, wind energy cost a fifth of what it did in the 1980s, and some expected that downward trend to continue as larger multi-megawatt turbines were mass-produced. As of 2012 capital costs for wind turbines are substantially lower than 2008–2010 but are still above 2002 levels. A 2011 report from the American Wind Energy Association stated, "Wind's costs have dropped over the past two years, in the range of 5 to 6 cents per kilowatt-hour recently.... about 2 cents cheaper than coal-fired electricity, and more projects were financed through debt arrangements than tax equity structures last year.... winning more mainstream acceptance from Wall Street's banks.... Equipment makers can also deliver products in the same year that they are ordered instead of waiting up to three years as was the case in previous cycles.... 5,600 MW of new installed capacity is under construction in the United States, more than double the number at this point in 2010. Thirty-five percent of all new power generation built in the United States since 2005 has come from wind, more than new gas and coal plants combined, as power providers are increasingly enticed to wind as a convenient hedge against unpredictable commodity price moves."

A British Wind Energy Association report gives an average generation cost of onshore wind power of around 3.2 pence (between US 5 and 6 cents) per kW ·h (2005). Cost per unit of energy produced was estimated in 2006 to be comparable to the cost of new generating capacity in the US for coal and natural gas: wind cost was estimated at $55.80 per MW ·h, coal at $53.10/MW ·h and natural gas at $52.50. Similar comparative results with natural gas were obtained in a governmental study in the UK in 2011. The presence of wind energy, even when subsidised, can reduce costs for consumers (€5 billion/yr in Germany) by reducing the marginal price, by minimising the use of expensive peaking power plants.

In February 2013 Bloomberg New Energy Finance reported that the cost of generating electricity from new wind farms is cheaper than new coal or new baseload gas plants. When including the current Australian federal government carbon pricing scheme their modelling gives costs (in Australian dollars) of $80/MWh for new wind farms, $143/MWh for new coal plants and $116/MWh for new baseload gas plants. The modelling also shows that "even without a carbon price (the most efficient way to reduce economy-wide emissions) wind energy is 14% cheaper than new coal and 18% cheaper than new gas." Part of the higher costs for new coal plants is due to high financial lending costs because of "the reputational damage of emissions-intensive investments". The expense of gas fired plants is partly due to "export market" effects on local prices. Costs of production from coal fired plants built in "the 1970s and 1980s" are cheaper than renewable energy sources because of depreciation.

An EU study shows base cost of onshore wind power similar to coal, when subsidies and externalities are disregarded. Wind power has some of the lowest external costs.

This cost has additionally reduced as wind turbine technology has improved. There are now longer and lighter wind turbine blades, improvements in turbine performance and increased power generation efficiency. Also, wind project capital and maintenance costs have continued to decline. For example, the wind industry in the USA is now able to produce more power at lower cost by using taller wind turbines with longer blades, capturing the faster winds at higher elevations. This has opened up new opportunities and in Indiana, Michigan, and Ohio, the price of power from wind turbines built 300 feet to 400 feet above the ground can now compete with conventional fossil fuels like coal. Prices have fallen to about 4 cents per kilowatt-hour in some cases and utilities have been increasing the amount of wind energy in their portfolio, saying it is their cheapest option.

Incentives and Community Benefits

The U.S. wind industry generates tens of thousands of jobs and billions of dollars of economic activity. Wind projects provide local taxes, or payments in lieu of taxes and strengthen the economy of rural communities by providing income to farmers with wind turbines on their land. Wind energy in many jurisdictions receives financial or other support to encourage its development. Wind energy benefits from subsidies in many jurisdictions, either to increase its attractiveness, or to compensate for subsidies received by other forms of production which have significant negative externalities.

In the US, wind power receives a production tax credit (PTC) of 1.5¢/kWh in 1993 dollars for each kW·h produced, for the first ten years; at 2.2 cents per kW·h in 2012, the credit was renewed on 2 January 2012, to include construction begun in 2013. A 30% tax credit can be applied instead of receiving the PTC. Another tax benefit is accelerated depreciation. Many American states also provide incentives, such as exemption from property tax, mandated purchases, and additional markets for "green credits". The Energy Improvement and Extension Act of 2008 contains extensions of credits for wind, including microturbines. Countries such as Canada and Germany also provide incentives for wind turbine construction, such as tax credits or minimum purchase prices for wind generation, with assured grid access (sometimes referred to as feed-in tariffs). These feed-in tariffs are typically set well above average electricity prices. In December 2013 U.S. Senator Lamar Alexander and other Republican senators argued that the "wind energy production tax credit should be allowed to expire at the end of 2013" and it expired 1 January 2014 for new installations.

Secondary market forces also provide incentives for businesses to use wind-generated power, even if there is a premium price for the electricity. For example, socially responsible manufacturers pay utility companies a premium that goes to subsidize and build new wind power infrastructure. Companies use wind-generated power, and in return they can claim that they are undertaking strong "green" efforts. In the US the organisation Green-e monitors business compliance with these renewable energy credits.

Small-Scale Wind Power

Small-scale wind power is the name given to wind generation systems with the capacity to produce up to 50 kW of electrical power. Isolated communities, that may otherwise rely on diesel generators, may use wind turbines as an alternative. Individuals may purchase these systems to reduce or eliminate their dependence on grid electricity for economic reasons, or to reduce their carbon footprint. Wind turbines have been used for household electricity generation in conjunction with battery storage over many decades in remote areas.

Recent examples of small-scale wind power projects in an urban setting can be found in New York City, where, since 2009, a number of building projects have capped their roofs with Gorlov-type helical wind turbines. Although the energy they generate is small compared to the buildings' overall consumption, they help to reinforce the building's 'green' credentials in ways that "showing people your high-tech boiler" can not, with some of the projects also receiving the direct support of the New York State Energy Research and Development Authority.

Grid-connected domestic wind turbines may use grid energy storage, thus replacing purchased electricity with locally produced power when available. The surplus power produced by domestic microgenerators can, in some jurisdictions, be fed into the network and sold to the utility company, producing a retail credit for the microgenerators' owners to offset their energy costs.

Off-grid system users can either adapt to intermittent power or use batteries, photovoltaic or diesel systems to supplement the wind turbine. Equipment such as parking meters, traffic warning signs, street lighting, or wireless Internet gateways may be powered by a small wind turbine, possibly combined with a photovoltaic system, that charges a small battery replacing the need for a connection to the power grid.

A Carbon Trust study into the potential of small-scale wind energy in the UK, published in 2010, found that small wind turbines could provide up to 1.5 terawatt hours (TW·h) per year of electricity (0.4% of total UK electricity consumption), saving 0.6 million tonnes of carbon dioxide (Mt CO_2) emission savings. This is based on the assumption that 10% of households would install turbines at costs competitive with grid electricity, around 12 pence (US 19 cents) a kW·h. A report prepared for the UK's government-sponsored Energy Saving Trust in 2006, found that home power generators of various kinds could provide 30 to 40% of the country's electricity needs by 2050.

Distributed generation from renewable resources is increasing as a consequence of the increased awareness of climate change. The electronic interfaces required to connect renewable generation units with the utility system can include additional functions, such as the active filtering to enhance the power quality.

Environmental Effects

Figure: *Livestock grazing near a wind turbine.*

The environmental impact of wind power when compared to the environmental impacts of fossil fuels, is relatively minor. According to the IPCC, in assessments of the life-cycle global warming potential of energy sources, wind turbines have a median value of between 12 and 11 (gCO2eq/kWh) depending, respectively, on if offshore or onshore turbines are being assessed. Compared with other low carbon power sources, wind turbines have some of the lowest global warming potential per unit of electrical energy generated.

While a wind farm may cover a large area of land, many land uses such as agriculture are compatible with it, as only small areas of turbine foundations and infrastructure are made unavailable for use. There are reports of bird and bat mortality at wind turbines as there are around other artificial structures. The scale of the ecological impact may or may not be significant, depending on specific circumstances. Prevention and mitigation of wildlife fatalities, and protection of peat bogs, affect the siting and operation of wind turbines.

There are anecdotal reports of negative health effects from noise on people who live very close to wind turbines. Peer-reviewed research has generally not supported these claims. Aesthetic aspects of wind turbines and resulting changes of the visual landscape are significant. Conflicts arise especially in scenic and heritage protected landscapes.

Wind Turbine

A wind turbine is a device that converts kinetic energy from the wind into electrical power. A wind turbine used for charging batteries may be referred to as a wind charger.

The result of over a millennium of windmill development and modern engineering, today's wind turbines are manufactured in a wide range of vertical and horizontal axis types. The smallest turbines are used for applications such as battery charging for auxiliary power for boats or caravans or to power traffic warning signs. Slightly larger turbines can be used for making contributions to a domestic power supply while selling unused power back to the utility supplier via the electrical grid. Arrays of large turbines, known as wind farms, are becoming an increasingly important source of renewable energy and are used by many countries as part of a strategy to reduce their reliance on fossil fuels.

History

Windmills were used in Persia (present-day Iran) as early as 200 B.C. The windwheel of Hero of Alexandria marks one of the first known instances of wind powering a machine in history. However, the first known practical windmills were built in Sistan, an Eastern province of Iran, from the 7th century. These "Panemone" were vertical axle windmills, which had long vertical drive shafts with rectangular blades. Made of six to twelve sails covered in reed matting or cloth material, these windmills were used to grind grain or draw up water, and were used in the gristmilling and sugarcane industries.

Figure: *James Blyth's electricity-generating wind turbine, photographed in 1891*

Windmills first appeared in Europe during the Middle Ages. The first historical records of their use in England date to the 11th or 12th centuries and there are reports of German crusaders taking their windmill-making skills to Syria around 1190. By the 14th century, Dutch windmills were in use to drain areas of the Rhine delta.

The first electricity-generating wind turbine was a battery charging machine installed in July 1887 by Scottish academic James Blyth to light his holiday home in Marykirk, Scotland. Some months later American inventor Charles F. Brush built the first automatically operated wind turbine for electricity production in Cleveland, Ohio. Although Blyth's turbine was considered uneconomical in the United Kingdom electricity generation by wind turbines was more cost effective in countries with widely scattered populations.

In Denmark by 1900, there were about 2500 windmills for mechanical loads such as pumps and mills, producing an estimated combined peak power of about 30 MW. The largest machines were on 24-meter (79 ft) towers with four-bladed 23-meter (75 ft) diameter rotors. By 1908 there were 72 wind-driven electric generators operating in the United States from 5 kW to 25 kW. Around the time of World War I, American windmill makers were producing 100,000 farm windmills each year, mostly for water-pumping.

By the 1930s, wind generators for electricity were common on farms, mostly in the United States where distribution systems had not yet been installed. In this period, high-tensile steel was cheap, and the generators were placed atop prefabricated open steel lattice towers.

A forerunner of modern horizontal-axis wind generators was in service at Yalta, USSR in 1931. This was a 100 kW generator on a 30-meter (98 ft) tower, connected to the local 6.3 kV distribution system. It was reported to have an annual capacity factor of 32 percent, not much different from current wind machines.

In the autumn of 1941, the first megawatt-class wind turbine was synchronized to a utility grid in Vermont. The Smith-Putnam wind turbine only ran for 1,100 hours before suffering a critical failure. The unit was not repaired, because of shortage of materials during the war.

The first utility grid-connected wind turbine to operate in the UK was built by John Brown & Company in 1951 in the Orkney Islands.

Despite these diverse developments, developments in fossil fuel systems almost entirely eliminated any wind turbine systems larger than supermicro size. In the early 1970s, however, anti-nuclear protests in Denmark spurred artisan mechanics to develop microturbines of 22 kW. Organising owners into associations and co-operatives lead to the lobbying of the government and utilities and provided incentives for larger turbines throughout the 1980s and later. Local activists in Germany, nascent turbine manufacturers in Spain, and large investors in the United States in the early 1990s then lobbied for policies that stimulated the industry in those countries. Later companies formed in India and China. As of 2012, Danish company Vestas is the world's biggest wind-turbine manufacturer.

Resources

A quantitative measure of the wind energy available at any location is called the Wind Power Density (WPD) It is a calculation of the mean annual power available per square meter of swept area of a turbine, and is tabulated for different heights above ground. Calculation of wind power density includes the effect of wind velocity and air density. Colour-coded maps are prepared for a particular area described, for example, as "Mean Annual Power Density at 50 Metres". In the United States, the results of the above calculation are included in an index developed by the National Renewable Energy Laboratory and referred to as "NREL CLASS". The larger the WPD calculation, the higher it is rated by class. Classes range from Class 1 (200 watts per square meter or less at 50 m altitude) to Class 7 (800 to 2000 watts per square m).

Commercial wind farms generally are sited in Class 3 or higher areas, although isolated points in an otherwise Class 1 area may be practical to exploit.

Wind turbines are classified by the wind speed they are designed for, from class I to class IV, with A or B referring to the turbulence.

Class	*Avg Wind Speed (m/s)*	*Turbulence*
IA	10	18%
IB	10	16%
IIA	8.5	18%
IIB	8.5	16%
IIIA	7.5	18%
IIIB	7.5	16%
IVA	6	18%
IVB	6	16%

Efficiency

Not all the energy of blowing wind can be harvested, since conservation of mass requires that as much mass of air exits the turbine as enters it. Betz's law gives the maximal achievable extraction of wind power by a wind turbine as 59% of the total kinetic energy of the air flowing through the turbine.

Further inefficiencies, such as rotor blade friction and drag, gearbox losses, generator and converter losses, reduce the power delivered by a wind turbine. Commercial utility-connected turbines deliver 75% to 80% of the Betz limit of power extractable from the wind, at rated operating speed.

Efficiency can decrease slightly over time due to wear. Analysis of 3128 wind turbines older than 10 years in Denmark showed that half of the turbines had no decrease, while the other half saw a production decrease of 1.2% per year.

Types

Wind turbines can rotate about either a horizontal or a vertical axis, the former being both older and more common.

Horizontal Axis

Horizontal-axis wind turbines (HAWT) have the main rotor shaft and electrical generator at the top of a tower, and must be pointed into the wind. Small turbines are pointed by a simple wind vane, while large turbines generally use a wind sensor coupled with a servo motor.

Most have a gearbox, which turns the slow rotation of the blades into a quicker rotation that is more suitable to drive an electrical generator.

Figure: *Components of a horizontal axis wind turbine (gearbox, rotor shaft and brake assembly) being lifted into position*

Figure: *A turbine blade convoy passing through Edenfield, UK*

Since a tower produces turbulence behind it, the turbine is usually positioned upwind of its supporting tower. Turbine blades are made stiff to prevent the blades from being pushed into the tower by high winds. Additionally, the blades are placed a considerable distance in front of the tower and are sometimes tilted forward into the wind a small amount.

Downwind machines have been built, despite the problem of turbulence (mast wake), because they don't need an additional mechanism for keeping them in line with the wind, and because in high winds the blades can be allowed to bend which reduces their swept area and thus their wind resistance. Since cyclical (that is repetitive) turbulence may lead to fatigue failures, most HAWTs are of upwind design.

Turbines used in wind farms for commercial production of electric power are usually three-bladed and pointed into the wind by computer-controlled motors. These have high tip speeds of over 320 km/h (200 mph), high efficiency, and low torque ripple, which contribute to good reliability. The blades are usually coloured white for daytime visibility by aircraft and range in length from 20 to 40 meters (66 to 131 ft) or more. The tubular steel towers range from 60 to 90 meters (200 to 300 ft) tall. The blades rotate at 10 to 22 revolutions per minute. At 22 rotations per minute the tip speed exceeds 90 meters per second (300 ft/s). A gear box is commonly used for stepping up the speed of the generator, although designs may also use direct drive of an annular generator. Some models operate at constant speed, but more energy can be collected by variable-speed turbines which use a solid-state power converter to interface to the transmission system. All turbines are equipped with protective features to avoid damage at high wind speeds, by feathering the blades into the wind which ceases their rotation, supplemented by brakes.

Vertical Axis Design

Vertical-axis wind turbines (or VAWTs) have the main rotor shaft arranged vertically. One advantage of this arrangement is that the turbine does not need to be pointed into the wind to be effective, which is an advantage on a site where the wind direction is highly variable. It is also an advantage when the turbine is integrated into a building because it is inherently less steerable. Also, the generator and gearbox can be placed near the ground, using a direct drive from the rotor assembly to the ground-based gearbox, improving accessibility for maintenance.

The key disadvantages include the relatively low rotational speed with the consequential higher torque and hence higher cost of the drive train, the inherently lower power coefficient, the 360 degree rotation of the aerofoil within the wind flow during each cycle and hence the highly dynamic loading on the blade, the pulsating torque generated by some rotor designs on the drive train, and the difficulty of modelling the wind flow accurately and hence the challenges of analysing and designing the rotor prior to fabricating a prototype.

When a turbine is mounted on a rooftop the building generally redirects wind over the roof and this can double the wind speed at the turbine. If the height of a rooftop mounted turbine tower is approximately 50% of the building height it is near the optimum for maximum wind energy and minimum wind turbulence. Wind speeds within the built environment are generally much lower than at exposed rural sites, noise may be a concern and an existing structure may not adequately resist the additional stress.

Subtypes of the Vertical Axis Design Include

Figure: *Offshore Horizontal Axis Wind Turbines (HAWTs) at Scroby Sands Wind Farm, UK*

Darrieus Wind Turbine

"Eggbeater" turbines, or Darrieus turbines, were named after the French inventor, Georges Darrieus. They have good efficiency, but produce large torque ripple and cyclical stress on the tower, which contributes to poor reliability. They also generally require some external power source, or an additional Savonius rotor to start turning, because the starting torque is very low. The torque ripple is reduced by using three or more blades which results in greater solidity of the rotor. Solidity is measured by blade area divided by the rotor area. Newer Darrieus type turbines are not held up by guy-wires but have an external superstructure connected to the top bearing.

Giromill

A subtype of Darrieus turbine with straight, as opposed to curved, blades. The cycloturbine variety has variable pitch to reduce the torque pulsation and is self-starting. The advantages of variable pitch are: high starting torque; a wide, relatively flat torque curve; a higher coefficient of performance; more efficient operation in turbulent winds; and a lower blade speed ratio which lowers blade bending stresses. Straight, V, or curved blades may be used.

Savonius Wind Turbine

These are drag-type devices with two (or more) scoops that are used in anemometers, *Flettner* vents (commonly seen on bus and van roofs), and in some high-reliability low-efficiency power turbines. They are always self-starting if there are at least three scoops.

Twisted Savonius

Twisted Savonius is a modified savonius, with long helical scoops to provide smooth torque. This is often used as a rooftop windturbine and has even been adapted for ships. Another type of vertical axis is the Parallel turbine, which is similar to the crossflow fan or centrifugal fan. It uses the ground effect. Vertical axis turbines of this type have been tried for many years: a unit producing 10 kW was built by Israeli wind pioneer Bruce Brill in the 1980s.

Wind Turbine Design

Wind turbine design is the process of defining the form and specifications of a wind turbine to extract energy from the wind. A wind turbine installation consists of the necessary systems needed to capture the wind's energy, point the turbine into the wind, convert mechanical rotation into electrical power, and other systems to start, stop, and control the turbine. This chapter covers the design of horizontal axis wind turbines (HAWT) since the majority of commercial turbines use this design. In 1919 the physicist Albert Betz showed that for a hypothetical ideal wind-energy extraction machine, the fundamental laws of conservation of mass and energy allowed no more than 16/27 (59.3%) of the kinetic energy of the wind to be captured. This Betz' law limit can be approached by modern turbine designs which may reach 70 to 80% of this theoretical limit. In addition to aerodynamic design of the blades, design of a complete wind power system must also address design of the hub, controls, generator, supporting structure and foundation. Further design questions arise when integrating wind turbines into electrical power grids.

Aerodynamics

The shape and dimensions of the blades of the wind turbine are determined by the aerodynamic performance required to efficiently extract energy from the wind, and by the strength required to resist the forces on the blade.

The aerodynamics of a horizontal-axis wind turbine are not straightforward. The air flow at the blades is not the same as the airflow far away from the turbine. The very nature of the way in which energy is extracted from the air also causes air to be deflected by the turbine. In addition the aerodynamics of a wind turbine at the rotor surface exhibit phenomena that are rarely seen in other aerodynamic fields.

In 1919 the physicist Albert Betz showed that for a hypothetical ideal wind-energy extraction machine, the fundamental laws of conservation of mass and energy allowed no more than 16/27 (59.3%) of the kinetic energy of the wind to be captured. This Betz' law limit can be approached by modern turbine designs which may reach 70 to 80% of this theoretical limit.

Power Control

The speed at which a wind turbine rotates must be controlled for efficient power generation and to keep the turbine components within designed speed and torque limits. The centrifugal force on the spinning blades increases as the square of the rotation speed, which makes this structure sensitive to overspeed. Because the power of the wind increases as the cube of the wind speed, turbines have to be built to survive much higher wind loads (such as gusts of wind) than those from which they can practically generate power. Wind turbines have ways of reducing torque in high winds.

A wind turbine is designed to produce power over a range of wind speeds. All wind turbines are designed for a maximum wind speed, called the survival speed, above which they will be damaged. The survival speed of commercial wind turbines is in the range of 40 m/s (144 km/h, 89 MPH) to 72 m/s (259 km/h, 161 MPH). The most common survival speed is 60 m/s (216 km/h, 134 MPH).

If the rated wind speed is exceeded the power has to be limited. There are various ways to achieve this. A control system involves three basic elements: sensors to measure process variables, actuators to manipulate energy capture and component loading, and control algorithms to coordinate the actuators based on information gathered by the sensors.

Stall

Stalling works by increasing the angle at which the relative wind strikes the blades (angle of attack), and it reduces the induced drag (drag associated with lift). Stalling is simple because it can be made to happen passively (it increases automatically when the winds speed up), but it increases the cross-section of the blade face-on to the wind, and thus the ordinary drag. A fully stalled turbine blade, when stopped, has the flat side of the blade facing directly into the wind.

A fixed-speed HAWT inherently increases its angle of attack at higher wind speed as the blades speed up. A natural strategy, then, is to allow the blade to stall when the wind speed increases. This technique was successfully used on many early HAWTs. However, on some of these blade sets, it was observed that the degree of blade pitch tended to increase audible noise levels. Vortex generators may be used to control the lift characteristics of the blade. The VGs are placed on the airfoil to enhance the lift if they are placed on the lower (flatter) surface or limit the maximum lift if placed on the upper (higher camber) surface.

Pitch Control

Furling works by decreasing the angle of attack, which reduces the induced drag from the lift of the rotor, as well as the cross-section. One major problem in designing wind turbines is getting the blades to stall or furl quickly enough should a gust of wind cause sudden acceleration. A fully furled turbine blade, when stopped, has the edge of the blade facing into the wind.

Loads can be reduced by making a structural system softer or more flexible. This could be accomplished with downwind rotors or with curved blades that twist naturally to reduce angle of attack at higher wind speeds. These systems will be nonlinear and will couple the structure to the flow field - thus, design tools must evolve to model these nonlinearities. Standard modern turbines all furl the blades in high winds. Since furling requires acting against the torque on the blade, it requires some form of pitch angle control, which is achieved with a slewing drive. This drive precisely angles the blade while withstanding high torque loads. In addition, many turbines use hydraulic systems. These systems are usually spring-loaded, so that if hydraulic power fails, the blades automatically furl. Other turbines use an electric servomotor for every rotor blade. They have a small battery-reserve in case of an electric-grid breakdown. Small wind turbines (under 50 kW) with variable-pitching generally use systems operated by centrifugal force, either by flyweights or geometric design, and employ no electric or hydraulic controls.

Fundamental gaps exist in pitch control, limiting the reduction of energy costs, according to a report from a coalition of researchers from universities, industry, and government, supported by the Atkinson Centre for a Sustainable Future. Load reduction is currently focused on full-span blade pitch control, since individual pitch motors are the actuators currently available on commercial turbines. Significant load mitigation has been demonstrated in simulations for blades, tower, and drive train. However, there is still research needed, the methods for realisation of full-span blade pitch control need to be developed in order to increase energy capture and mitigate fatigue loads.

A control technique applied to the pitch angle is done by comparing the current active power of the engine with the value of active power at the rated engine speed (active power reference, Ps reference). Control of the pitch angle in this case is done with a PI controller controls. However, in order to have a realistic response to the control system of the pitch angle, the actuator uses the time constant Tservo, an integrator and limiters so as the pitch angle to be from 0° to 30° with a rate of change (± 10° per sec).

From the figure at the right, the reference pitch angle is compared with the actual pitch angle b and then the error is corrected by the actuator. The reference pitch angle, which comes from the PI controller, goes through a limiter. Restrictions on limits are very important to maintain the pitch angle in real terms. Limiting the rate of change is very important especially during faults in the network. The importance is due to the fact that the controller decides how quickly it can reduce the aerodynamic energy to avoid acceleration during errors.

Other Controls

Generator Torque: Modern large wind turbines are variable-speed machines. When the wind speed is below rated, generator torque is used to control the rotor speed in order to capture as much power as possible. The most power is captured when the tip speed ratio is held constant at its optimum value (typically 6 or 7). This means that as wind speed increases, rotor speed should increase proportionally. The difference between the aerodynamic torque captured by the blades and the applied generator torque controls the rotor speed. If the generator torque is lower, the rotor accelerates, and if the generator torque is higher, the rotor slows down. Below rated wind speed, the generator torque control is active while the blade pitch is typically held at the constant angle that captures the most power, fairly flat to the wind. Above rated wind speed, the generator torque is typically held constant while the blade pitch is active. One technique to control a permanent magnet synchronous motor is Field Oriented Control. Field Oriented Control is a closed loop strategy composed of two current controllers (an inner loop and outer loop cascade design) necessary for controlling the torque, and one speed controller.

Constant Torque Angle Control

In this control strategy the d axis current is kept zero, while the vector current is align with the q axis in order to maintain the torque angle equal with 90°. This is one of the most used control strategy because of the simplicity, by controlling only the Iqs current. So, now the electromagnetic torque equation of the permanent magnet synchronous generator is simply a linear equation depend on the Iqs current only.

So, the electromagnetic torque for Ids = 0 (we can achieve that with the d-axis controller) is now:

$$T_e = 3/2\ p\ (\lambda_{pm}\ I_{qs} + (L_{ds}-L_{qs})\ I_{ds}\ I_{qs}) = 3/2\ p\ \lambda_{pm}\ I_{qs}$$

So, the complete system of the machine side converter and the cascaded PI controller loops is given by the figure in the right. In that we have the control inputs, which are the duty rations m_{ds} and m_{qs}, of the PWM-regulated converter. Also, we can see the control scheme for the wind turbine in the machine side and simultaneously how we keep the I_{ds} zero (the electromagnetic torque equation is linear).

Yawing

Modern large wind turbines are typically actively controlled to face the wind direction measured by a wind vane situated on the back of the nacelle. By minimizing the yaw angle (the misalignment between wind and turbine pointing direction), the power output is maximized and non-symmetrical loads minimized. However, since the wind direction varies quickly the turbine will not strictly follow the direction and will have a small yaw angle on average. The power output losses can simply be approximated to fall with (cos(yaw angle)). Particularly at low-to-medium wind speeds, yawing can make a significant reduction in turbine output, with wind direction variations of ±30° being quite common and long response times of the turbines to changes in wind direction. At high wind speeds, the wind direction is less variable.

Electrical Braking

Braking of a small wind turbine can be done by dumping energy from the generator into a resistor bank, converting the kinetic energy of the turbine rotation into heat. This method is useful if the kinetic load on the generator is suddenly reduced or is too small to keep the turbine speed within its allowed limit.

Cyclically braking causes the blades to slow down, which increases the stalling effect, reducing the efficiency of the blades. This way, the turbine's rotation can be kept at a safe speed in faster winds while maintaining (nominal) power output. This method is usually not applied on large grid-connected wind turbines.

Mechanical Braking

A mechanical drum brake or disk brake is used to stop turbine in emergency situation such as extreme gust events or over speed. This brake is a secondary means to hold the turbine at rest for maintenance, with a rotor lock system as primary means. Such brakes are usually applied only after blade furling and electromagnetic braking have reduced the turbine speed generally 1 or 2 rotor RPM, as the mechanical brakes can create a fire inside the nacelle if used to stop the turbine from full speed. The load on the turbine increases if the brake is applied at rated RPM. Mechanical brakes are driven by hydraulic systems and are connected to main control box.

Turbine Size

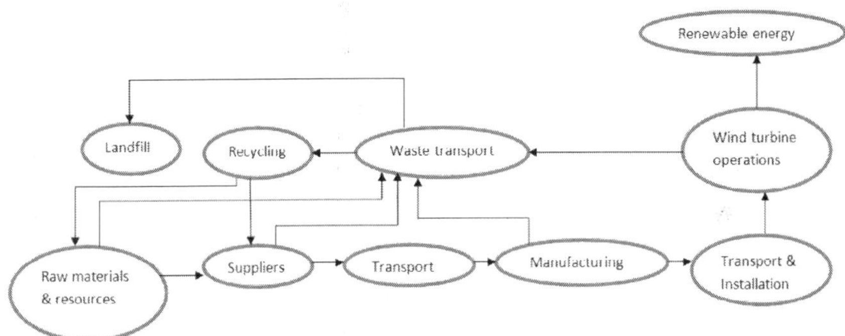

Figure: *Flow diagram for wind turbine plant*

There are different size classes of wind turbines. The smallest having power production less than 10 kW are used in homes, farms and remote applications whereas intermediate wind turbines (10-250 kW) are useful for village power, hybrid systems and distributed power. The largest wind turbines (660 kW – 2+MW) are used in central station wind farms, distributed power and community wind. For a given survivable wind speed, the mass of a turbine is approximately proportional to the cube of its blade-length. Wind power intercepted by the turbine is proportional to the square of its blade-length. The maximum blade-length of a turbine is limited by both the strength and stiffness of its material.

Labour and maintenance costs increase only gradually with increasing turbine size, so to minimize costs, wind farm turbines are basically limited by the strength of materials, and siting requirements. Typical modern wind turbines have diameters of 40 to 90 metres (130 to 300 ft) and are rated between 500 kW and 2 MW. As of 2014 the most powerful turbine, the Vestas V-164, is rated at 8 MW and has a rotor diameter of 164m.

Generator

Commercial size generators have a rotor carrying a field winding so that a rotating magnetic field is produced inside a set of windings called the stator. While the rotating field winding consumes a fraction of a percent of the generator output, adjustment of the field current allows good control over the generator output voltage.

For large, commercial size horizontal-axis wind turbines, the electrical generator is mounted in a nacelle at the top of a tower, behind the hub of the turbine rotor. Typically wind turbines generate electricity through asynchronous machines that are directly connected with the electricity grid. Usually the rotational speed of the wind turbine is slower than the equivalent rotation speed of the electrical network—typical rotation speeds for wind generators are 5–20 rpm while a directly connected machine will have an electrical speed between 750-3600 rpm. Therefore, a gearbox is inserted between the rotor hub and the generator. This also reduces the generator cost and weight.

In conventional wind turbines, the blades spin a shaft that is connected through a gearbox to the generator. The gearbox converts the turning speed of the blades 15 to 20 rotations per minute for a large, one-megawatt turbine into the faster 1,800 rotations per minute that the generator needs to generate electricity. Analysts from GlobalData estimate that gearbox market grows from $3.2bn in 2006 to $6.9bn in 2011, and to $8.1bn by 2020. Market leaders were Winergy in 2011. The use of magnetic gearboxes has also been explored as a way of reducing wind turbine maintenance costs.

Older style wind generators rotate at a constant speed, to match power line frequency, which allowed the use of less costly induction generators. Newer wind turbines often turn at whatever speed generates electricity most efficiently. The varying output frequency and voltage can be matched to the fixed values of the grid using multiple technologies such as doubly fed induction generators or full-effect converters where the variable frequency current produced is converted to DC and then back to AC. Although such alternatives require costly equipment and cause power loss, the turbine can capture a significantly larger fraction of the wind energy. In some cases, especially when turbines are sited offshore, the DC energy will be transmitted from the turbine to a central (onshore) inverter for connection to the grid.

Gearless wind turbines (also called direct drive) get rid of the gearbox completely. Instead, the rotor shaft is attached directly to the generator, which spins at the same speed as the blades. Enercon and EWT (Formerly known as Lagerwey) have produced gearless wind turbines with separately electrically excited generators for many years, and Siemens produces a gearless "inverted generator" 3 MW model while developing a 6 MW model. To make up for a direct drive generator's slower spinning rate, the diameter of the generator's rotor is increased hence containing more magnets which lets it create a lot of power when turning slowly.

Gearless wind turbines are often heavier than gear based wind turbines. A study by the EU called "Reliawind" based on the largest sample size of turbines, has shown that the reliability of gearboxes is not the main problem in wind turbines. The reliability of direct drive turbines offshore is still not known, since the sample size is so small.

Experts from Technical University of Denmark estimate that a geared generator with permanent magnets may use 25 kg/MW of the rare earth element Neodymium, while a gearless may use 250 kg/MW.

In December 2011, the US Department of Energy published a report stating critical shortage of rare earth elements such as neodymium used in large quantities for permanent magnets in gearless wind turbines. China produces more than 95% of rare earth elements, while Hitachi holds more than 600 patents covering Neodymium magnets. Direct-drive turbines require 600 kg of PM material per megawatt, which translates to several hundred kilograms of rare earth content per megawatt, as neodymium content is estimated to be 31% of magnet weight. Hybrid drivetrains (intermediate between direct drive and traditional geared) use significantly less rare earth materials. While PM wind turbines only account for about 5% of the market outside of China, their market share inside of China is estimated at 25% or higher. Demand for neodymium in wind turbines is estimated to be 1/5 of that in electric vehicles.

Blades

Blade Design: The ratio between the speed of the blade tips and the speed of the wind is called tip speed ratio. High efficiency 3-blade-turbines have tip speed/wind speed ratios of 6 to 7. Modern wind turbines are designed to spin at varying speeds (a consequence of their generator design). Use of aluminum and composite materials in their blades has contributed to low rotational inertia, which means that newer wind turbines can accelerate quickly if the winds pick up, keeping the tip speed ratio more nearly constant. Operating closer to their optimal tip speed ratio during energetic gusts of wind allows wind turbines to improve energy capture from sudden gusts that are typical in urban settings.

In contrast, older style wind turbines were designed with heavier steel blades, which have higher inertia, and rotated at speeds governed by the AC frequency of the power lines. The high inertia buffered the changes in rotation speed and thus made power output more stable.

It is generally understood that noise increases with higher blade tip speeds. To increase tip speed without increasing noise would allow reduction the torque into the gearbox and generator and reduce overall structural loads, thereby reducing cost. The reduction of noise is linked to the detailed aerodynamics of the blades, especially factors that reduce abrupt stalling. The inability to predict stall restricts the development of aggressive aerodynamic concepts. A blade can have a lift-to-drag ratio of 120, compared to 70 for a sailplane and 15 for an airliner.

The Hub

In simple designs, the blades are directly bolted to the hub and hence are stalled. In other more sophisticated designs, they are bolted to the pitch mechanism, which adjusts their angle of attack according to the wind speed to control their rotational speed. The pitch mechanism is itself bolted to the hub. The hub is fixed to the rotor shaft which drives the generator directly or through a gearbox.

Blade Count

The number of blades is selected for aerodynamic efficiency, component costs, and system reliability. Noise emissions are affected by the location of the blades upwind or downwind of the tower and the speed of the rotor. Given that the noise emissions from the blades' trailing edges and tips vary by the 5th power of blade speed, a small increase in tip speed can make a large difference.

Wind turbines developed over the last 50 years have almost universally used either two or three blades. However, there are patents that present designs with additional blades, such as Chan Shin's Multi-unit rotor blade system integrated wind turbine. Aerodynamic efficiency increases with number of blades but with diminishing return. Increasing the number of blades from one to two yields a six percent increase in aerodynamic efficiency, whereas increasing the blade count from two to three yields only an additional three percent in efficiency. Further increasing the blade count yields minimal improvements in aerodynamic efficiency and sacrifices too much in blade stiffness as the blades become thinner.

Theoretically, an infinite number of blades of zero width is the most efficient, operating at a high value of the tip speed ratio. But other considerations lead to a compromise of only a few blades.

Component costs that are affected by blade count are primarily for materials and manufacturing of the turbine rotor and drive train. Generally, the fewer the number of blades, the lower the material and manufacturing costs will be. In addition, the fewer the number of blades, the higher the rotational speed can be. This is because blade stiffness requirements to avoid interference with the tower limit how thin the blades can be manufactured, but only for upwind machines; deflection of blades in a downwind machine results in increased tower clearance. Fewer blades with higher rotational speeds reduce peak torques in the drive train, resulting in lower gearbox and generator costs.

System reliability is affected by blade count primarily through the dynamic loading of the rotor into the drive train and tower systems. While aligning the wind turbine to changes in wind direction (yawing), each blade experiences a cyclic load at its root end depending on blade position. This is true of one, two, three blades or more. However, these cyclic loads when combined together at the drive train shaft are symmetrically balanced for three blades, yielding smoother operation during turbine yaw. Turbines with one or two blades can use a pivoting teetered hub to also nearly eliminate the cyclic loads into the drive shaft and system during yawing. A Chinese 3.6 MW two-blade is being tested in Denmark. Mingyang won a bid for 87 MW (29 * 3 MW) two-bladed offshore wind turbines near Zhuhai in 2013.

Finally, aesthetics can be considered a factor in that some people find that the three-bladed rotor is more pleasing to look at than a one- or two-bladed rotor.

Blade Materials

Wood and canvas sails were used on early windmills due to their low price, availability, and ease of manufacture. Smaller blades can be made from light metals such as aluminium. These materials, however, require frequent maintenance. Wood and canvas construction limits the airfoil shape to a flat plate, which has a relatively high ratio of drag to force captured (low aerodynamic efficiency) compared to solid airfoils. Construction of solid airfoil designs requires inflexible materials such as metals or composites. Some blades also have incorporated lightning conductors.

New wind turbine designs push power generation from the single megawatt range to upwards of 10 megawatts using larger and larger blades. A larger area effectively increases the tip-speed ratio of a turbine at a given wind speed, thus increasing its energy extraction. Computer-aided engineering software such as HyperSizer (originally developed for spacecraft design) can be used to improve blade design.

As of 2013, production wind turbine blades are as large as 120 meters in diameter with prototypes reaching 160 meters. In 2001, an estimated 50 million kilograms of fibreglass laminate were used in wind turbine blades.

An important goal of larger blade systems is to control blade weight. Since blade mass scales as the cube of the turbine radius, loading due to gravity constrains systems with larger blades. Gravitational loads include axial and tensile/ compressive loads (top/bottom of rotation) as well as bending (lateral positions). The magnitude of these loads fluctuates cyclically and the edgewise moments are reversed every 180° of rotation. Typical rotor speeds and design life are ~10rpm and 20 years, respectively, with the number of lifetime revolutions on the order of 10^8. Considering wind, it is expected that turbine blades go through ~10^9 loading cycles. Wind is another source of rotor blade loading. Lift causes bending in the flapwise direction (out of rotor plane) while air flow around the blade cause edgewise bending (in the rotor plane). Flapwise bending involves tension on the pressure (upwind) side and compression on the suction (downwind) side. Edgewise bending involves tension on the leading edge and compression on the trailing edge.

Wind loads are cyclical because of natural variability in wind speed and wind shear (higher speeds at top of rotation).

Failure in ultimate loading of wind-turbine rotor blades exposed to wind and gravity loading is a failure mode that needs to be considered when the rotor blades are designed. The wind speed that causes bending of the rotor blades exhibits a natural variability, and so does the stress response in the rotor blades. Also, the resistance of the rotor blades, in terms of their tensile strengths, exhibits a natural variability.

Manufacturing blades in the 40 to 50 metre range involves proven fibreglass composite fabrication techniques. Manufactures such as Nordex and GE Wind use an infusion process. Other manufacturers use variations on this technique, some including carbon and wood with fibreglass in an epoxy matrix. Other options include prepreg fibreglass and vacuum-assisted resin transfer molding. Each of these options use a glass-fibre reinforced polymer composite constructed with differing complexity. Perhaps the largest issue with more simplistic, open-mould, wet systems are the emissions associated with the volatile organics released. Preimpregnated materials and resin infusion techniques avoid the release of volatiles by containing all reaction gases. However, these contained processes have their own challenges, namely the production of thick laminates necessary for structural components becomes more difficult. As the preform resin permeability dictates the maximum laminate thickness, bleeding is required to eliminate voids and ensure proper resin distribution. One solution to resin distribution a partially preimpregnated fibreglass. During evacuation, the dry fabric provides a path for airflow and, once heat and pressure are applied, resin may flow into the dry region resulting in a thoroughly impregnated laminate structure.

Epoxy-based composites have environmental, production, and cost advantages over other resin systems. Epoxies also allow shorter cure cycles, increased durability, and improved surface finish. Prepreg operations further reduce processing time over wet lay-up systems. As turbine blades pass 60 metres, infusion techniques become more prevalent; the traditional resin transfer moulding injection time is too long as compared to the resin set-up time, limiting laminate thickness. Injection forces resin through a thicker ply stack, thus depositing the resin where in the laminate structure before gelatin occurs. Specialised epoxy resins have been developed to customize lifetimes and viscosity.

Carbon fibre-reinforced load-bearing spars can reduce weight and increase stiffness. Using carbon fibres in 60 metre turbine blades is estimated to reduce total blade mass by 38% and decrease cost by 14% compared to 100% fibreglass. Carbon fibres have the added benefit of reducing the thickness of fibreglass laminate sections, further addressing the problems associated with resin wetting of thick lay-up sections. Wind turbines may also benefit from the general trend of increasing use and decreasing cost of carbon fibre materials.

Tower

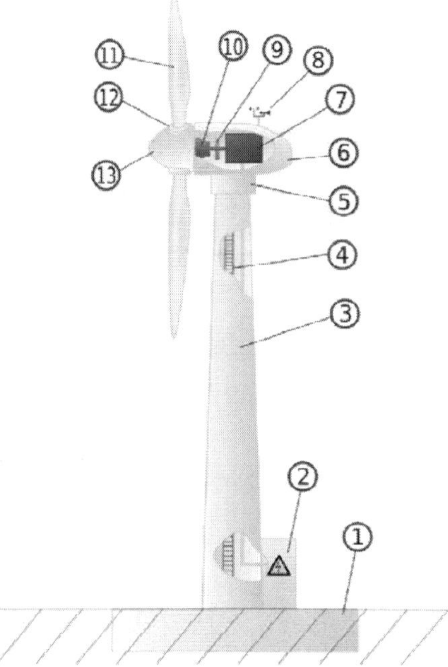

Figure: *Typical wind turbine components : 1-Foundation, 2-Connection to the electric grid, 3-Tower, 4-Access ladder, 5-Wind orientation control (Yaw control), 6-Nacelle, 7-Generator, 8-Anemometer, 9-Electric or Mechanical Brake, 10-Gearbox, 11-Rotor blade, 12-Blade pitch control, 13-Rotor hub.*

Typically, 2 types of towers exist: floating towers and land-based towers.

Tower Height

Wind velocities increase at higher altitudes due to surface aerodynamic drag (by land or water surfaces) and the viscosity of the air. The variation in velocity with altitude, called wind shear, is most dramatic near the surface.

Typically, in day the variation follows the wind profile power law, which predicts that wind speed rises proportionally to the seventh root of altitude. Doubling the altitude of a turbine, then, increases the expected wind speeds by 10% and the expected power by 34%. To avoid buckling, doubling the tower height generally requires doubling the diameter of the tower as well, increasing the amount of material by a factor of at least four.

At night time, or when the atmosphere becomes stable, wind speed close to the ground usually subsides whereas at turbine hub altitude it does not decrease that much or may even increase. As a result the wind speed is higher and a turbine will produce more power than expected from the 1/7 power law: doubling the altitude may increase wind speed by 20% to 60%. A stable atmosphere is caused by radiative cooling of the surface and is common in a temperate climate: it usually occurs when there is a (partly) clear sky at night. When the (high altitude) wind is strong (a 10-meter wind speed higher than approximately 6 to 7 m/s the stable atmosphere is disrupted because of friction turbulence and the atmosphere will turn neutral. A daytime atmosphere is either neutral (no net radiation; usually with strong winds and heavy clouding) or unstable (rising air because of ground heating—by the sun). Here again the 1/7 power law applies or is at least a good approximation of the wind profile. Indiana had been rated as having a wind capacity of 30,000 MW, but by raising the expected turbine height from 50 m to 70 m, the wind capacity estimate was raised to 40,000 MW, and could be double that at 100 m.

For HAWTs, tower heights approximately two to three times the blade length have been found to balance material costs of the tower against better utilisation of the more expensive active components.

Road size restrictions makes transportation of towers with a diameter of more than 4.3 m difficult. Swedish analyses show that it is important to have the bottom wing tip at least 30 m above the tree tops, but a taller tower requires a larger tower diameter. A 3 MW turbine may increase output from 5,000 MWh to 7,700 MWh per year by going from 80 to 125 meter tower height. A tower profile made of connected shells rather than cylinders can have a larger diameter and still be transportable. A 100 m prototype tower with TC bolted 18 mm 'plank' shells has been erected at the wind turbine test centree Høvsøre in Denmark and certified by Det Norske Veritas, with a Siemens nacelle. Shell elements can be shipped in standard 12 m shipping containers, and 2½ towers per week are produced this way.

Wood is being investigated as a material for wind turbine towers, and a 100 metre tall tower supporting a 1.5 MW turbine has been erected in Germany. The wood tower shares the same transportation benefits of the segmented steel shell tower, but without the steel resource consumption.

Connection to the Electric Grid

All grid-connected wind turbines, from the first one in 1939 until the development of variable-speed grid-connected wind turbines in the 1970s, were fixed-speed wind turbines. As of 2003, nearly all grid-connected wind turbines operate at exactly constant speed (synchronous generators) or within a few percent of constant speed (induction generators). As of 2011, many operational wind turbines used fixed speed induction generators (FSIG). As of 2011, most new grid-connected wind turbines are variable speed wind turbines—they are in some variable speed configuration.

Early wind turbine control systems were designed for peak power extraction, also called maximum power point tracking—they attempt to pull the maximum possible electrical power from a given wind turbine under the current wind conditions. More recent wind turbine control systems deliberately pull less electrical power than they possibly could in most circumstances, in order to provide other benefits, which include:

- spinning reserves to quickly produce more power when needed—such as when some other generator suddenly drops from the grid—up to the max power supported by the current wind conditions.
- Variable-speed wind turbines can (very briefly) produce more power than the current wind conditions can support, by storing some wind energy as kinetic energy (accelerating during brief gusts of faster wind) and later converting that kinetic energy to electric energy (decelerating, either when more power is needed elsewhere, or during short lulls in the wind, or both).
- damping (electrical) subsynchronous resonances in the grid
- damping (mechanical) resonances in the tower

The generator in a wind turbine produces alternating current (AC) electricity. Some turbines drive an AC/AC converter—which converts the AC to direct current (DC) with a rectifier and then back to AC with an inverter—in order to match the frequency and phase of the grid. However, the most common method in large modern turbines is to instead use a doubly fed induction generator directly connected to the electricity grid.

A useful technique to connect a permanent magnet synchronous generator to the grid is by using a back-to-back converter. Also, we can have control schemes so as to achieve unity power factor in the connection to the grid. In that way the wind turbine will not consume reactive power, which is the most common problem with wind turbines that use induction machines. This leads to a more stable power system. Moreover, with different control schemes a wind turbine with a permanent magnet synchronous generator can provide or consume reactive power. So, it can work as a dynamic capacitor/inductor bank so as to help with the power systems' stability.

Below we show the control scheme so as to achieve unity power factor:

Reactive power regulation consists of one PI controller in order to achieve operation with unity power factor (i.e. $Q_{grid} = 0$). It is obvious that I_{dN} has to be regulated to reach zero at steady-state ($I_{dNref} = 0$).

We can see the complete system of the grid side converter and the cascaded PI controller loops in the figure in the right.

Foundations

Wind turbines, by their nature, are very tall slender structures, this can cause a number of issues when the structural design of the foundations are considered.

The foundations for a conventional engineering structure are designed mainly to transfer the vertical load (dead weight) to the ground, this generally allows for a comparatively unsophisticated arrangement to be used. However in the case of wind turbines, due to the high wind and environmental loads experienced there is a significant horizontal dynamic load that needs to be appropriately restrained.

This loading regime causes large moment loads to be applied to the foundations of a wind turbine. As a result, considerable attention needs to be given when designing the footings to ensure that the turbines are sufficiently restrained to operate efficiently. In the current Det Norske Veritas (DNV) guidelines for the design of wind turbines the angular deflection of the foundations are limited to 0.5°. DNV guidelines regarding earthquakes suggest that horizontal loads are larger than vertical loads for offshore wind turbines, while guidelines for tsunamis only suggest designing for maximum sea waves. In contrast, IEC suggests considering tsunami loads.

Scale model tests using a 50-g centrifuge are being performed at the Technical University of Denmark to test monopile foundations for offshore wind turbines at 30 to 50-m water depth.

Costs

The modern wind turbine is a complex and integrated system. Structural elements comprise the majority of the weight and cost. All parts of the structure must be inexpensive, lightweight, durable, and manufacturable, under variable loading and environmental conditions. Turbine systems that have fewer failures, require less maintenance, are lighter and last longer will lead to reducing the cost of wind energy. One way to achieve this is to implement well-documented, validated analysis codes, according to a 2011 report from a coalition of researchers from universities, industry, and government, supported by the Atkinson Centree for a Sustainable Future. The major parts of a modern turbine may cost (percentage of total) : tower 22%, blades 18%, gearbox 14%, generator 8%.

Efficiency and Wind Speed

The efficiency of a wind turbine is a maximum at its design wind velocity, and efficiency decreases with the fluctuations in wind. The lowest velocity at which the turbine develops its full power is known as rated wind velocity. Below some minimum wind velocity, no useful power output can be produced from wind turbine. There are limits on both the minimum and maximum wind velocity for the efficient operation of wind turbines.

Design Specification

The design specification for a wind-turbine will contain a power curve and guaranteed availability. With the data from the wind resource assessment it is possible to calculate commercial viability. The typical operating temperature range is –20 to 40 °C (–4 to 104 °F). In areas with extreme climate (like Inner Mongolia or Rajasthan) specific cold and hot weather versions are required. Wind turbines can be designed and validated according to IEC 61400 standards.

Low Temperature

Utility-scale wind turbine generators have minimum temperature operating limits which apply in areas that experience temperatures below –20 °C. Wind turbines must be protected from ice accumulation. It can make anemometer readings inaccurate and which can cause high structure loads and damage. Some turbine manufacturers offer low-temperature packages at a few percent extra cost, which include internal heaters, different lubricants, and different alloys for structural elements. If the low-temperature interval is combined with a low-wind condition, the wind turbine will require an external supply of power, equivalent to a few percent of its rated power, for internal heating. For example, the St. Leon, Manitoba project has a total rating of 99 MW and is estimated to need up to 3 MW (around 3% of capacity) of station service power a few days a year for temperatures down to –30 °C. This factor affects the economics of wind turbine operation in cold climates.

Solar Energy

Solar power is the conversion of sunlight into electricity, either directly using photovoltaics (PV), or indirectly using concentrated solar power (CSP). Concentrated solar power systems use lenses or mirrors and tracking systems to focus a large area of sunlight into a small beam. Photovoltaics convert light into electric current using the photovoltaic effect.

Photovoltaics were initially, and still are, used to power small and medium-sized applications, from the calculator powered by a single solar cell to off-grid homes powered by a photovoltaic array. They are an important and relatively inexpensive source of electrical energy where grid power is inconvenient, unreasonably expensive to connect, or simply unavailable. However, as the cost of solar electricity is falling, solar power is also increasingly being used even in grid-connected situations as a way to feed low-carbon energy into the grid.

Commercial concentrated solar power plants were first developed in the 1980s. The 392 MW ISEGS CSP installation is the largest solar power plant in the world, located in the Mojave Desert of California. Other large CSP plants include the SEGS (354 MW) in the Mojave Desert of California, the Solnova Solar Power Station (150 MW) and the Andasol solar power station (150 MW), both in Spain. The 290 MW Agua Caliente Solar Project in the United States, and the 221 MW Charanka Solar Park in India, are the world's largest photovoltaic power stations.

Photovoltaics

Photovoltaics (PV) is a method of converting solar energy into direct current electricity using semiconducting materials that exhibit the photovoltaic effect. A photovoltaic system employs solar panels composed of a number of solar cells to supply usable solar power. Power generation from solar PV has long been seen as a clean sustainable energy technology which draws upon the planet's most plentiful and widely distributed renewable energy source – the sun. The direct conversion of sunlight to electricity occurs without any moving parts or environmental emissions during operation. It is well proven, as photovoltaic systems have now been used for fifty years in specialised applications, and grid-connected PV systems have been in use for over twenty years.

Driven by advances in technology and increases in manufacturing scale and sophistication, the cost of photovoltaics has declined steadily since the first solar cells were manufactured, and the levelised cost of electricity (LCOE) from PV is competitive with conventional electricity sources in an expanding list of geographic regions. Net metering and financial incentives, such as preferential feed-in tariffs for solar-generated electricity, have supported solar PV installations in many countries. With current technology, photovoltaics recoup the energy needed to manufacture them in 1.5 (in Southern Europe) to 2.5 years (in Northern Europe).

Solar PV is now, after hydro and wind power, the third most important renewable energy source in terms of globally installed capacity. More than 100 countries use solar PV. Installations may be ground-mounted (and sometimes integrated with farming and grazing) or built into the roof or walls of a building (either building-integrated photovoltaics or simply rooftop).

In 2013, the fast-growing capacity of worldwide installed solar PV increased by 38 percent to 139 gigawatts (GW). This is sufficient to generate at least 160 terawatt hours (TWh) or about 0.85 percent of the electricity demand on the planet. China, followed by Japan and the United States, is now the fastest growing market, while Germany remains the world's largest producer, contributing almost 6 percent to its national electricity demands.

Solar Cells

Photovoltaics are best known as a method for generating electric power by using solar cells to convert energy from the sun into a flow of electrons. The photovoltaic effect refers to photons of light exciting electrons into a higher state of energy, allowing them to act as charge carriers for an electric current. The photovoltaic effect was first observed by Alexandre-Edmond Becquerel in 1839. The term photovoltaic denotes the unbiased operating mode of a photodiode in which current through the device is entirely due to the transduced light energy. Virtually all photovoltaic devices are some type of photodiode. Solar cells produce direct current electricity from sun light which can be used to power equipment or to recharge a battery. The first practical application of photovoltaics was to power orbiting satellites and other spacecraft, but today the majority of photovoltaic modules are used for grid connected power generation. In this case an inverter is required to convert the DC to AC. There is a smaller market for off-grid power for remote dwellings, boats, recreational vehicles, electric cars, roadside emergency telephones, remote sensing, and cathodic protection of pipelines.

Photovoltaic power generation employs solar panels composed of a number of solar cells containing a photovoltaic material. Materials presently used for photovoltaics include monocrystalline silicon, polycrystalline silicon, amorphous silicon, cadmium telluride, and copper indium gallium selenide/sulfide. Copper solar cables connect modules (module cable), arrays (array cable), and sub-fields. Because of the growing demand for renewable energy sources, the manufacturing of solar cells and photovoltaic arrays has advanced considerably in recent years. Solar photovoltaics power generation has long been seen as a clean energy technology which draws upon the planet's most plentiful and widely distributed renewable energy source – the sun. The technology is "inherently elegant" in that the direct conversion of sunlight to electricity occurs without any moving parts or environmental emissions during operation. It is well proven, as photovoltaic systems have now been used for fifty years in specialised applications, and grid-connected systems have been in use for over twenty years.

Cells require protection from the environment and are usually packaged tightly behind a glass sheet. When more power is required than a single cell can deliver, cells are electrically connected together to form photovoltaic modules, or solar panels. A single module is enough to power an emergency telephone, but for a house or a power plant the modules must be arranged in multiples as arrays. Photovoltaic power capacity is measured as maximum power output under standardized test conditions (STC) in "Wp" (Watts peak). The actual power output at a particular point in time may be less than or greater than this standardized, or "rated," value, depending on geographical location, time of day, weather conditions, and other factors. Solar photovoltaic array capacity factors are typically under 25%, which is lower than many other industrial sources of electricity.

Current Developments

For best performance, terrestrial PV systems aim to maximize the time they face the sun. Solar trackers achieve this by moving PV panels to follow the sun. The increase can be by as much as 20% in winter and by as much as 50% in summer. Static mounted systems can be optimized by analysis of the sun path. Panels are often set to latitude tilt, an angle equal to the latitude, but performance can be improved by adjusting the angle for summer or winter. Generally, as with other semiconductor devices, temperatures above room temperature reduce the performance of photovoltaics.

A number of solar panels may also be mounted vertically above each other in a tower, if the zenith distance of the Sun is greater than zero, and the tower can be turned horizontally as a whole and each panels additionally around a horizontal axis. In such a tower the panels can follow the Sun exactly. Such a device may be described as a ladder mounted on a turnable disk. Each step of that ladder is the middle axis of a rectangular solar panel. In case the zenith distance of the Sun reaches zero, the "ladder" may be rotated to the north or the south to avoid a solar panel producing a shadow on a lower solar panel. Instead of an exactly vertical tower one can choose a tower with an axis directed to the polar star, meaning that it is parallel to the rotation axis of the Earth. In this case the angle between the axis and the Sun is always larger than 66 degrees. During a day it is only necessary to turn the panels around this axis to follow the Sun. Installations may be ground-mounted (and sometimes integrated with farming and grazing) or built into the roof or walls of a building (building-integrated photovoltaics).

Efficiency

The San Jose-based company Sunpower produces cells that have an energy conversion ratio of 19.5%, well above the market average of 12–18%. The most efficient solar cell so far is a multi-junction concentrator solar cell with an efficiency of 43.5% produced by Solar Junction in April 2011. The highest efficiencies achieved without concentration include Sharp Corporation at 35.8% using a proprietary triple-junction manufacturing technology in 2009, and Boeing Spectrolab (40.7% also using a triple-layer design).

Several companies have begun embedding power optimizers into PV modules called smart modules. These modules perform maximum power point tracking (MPPT) for each module individually, measure performance data for monitoring, and provide additional safety. Such modules can also compensate for shading effects, wherein a shadow falling across a section of a module causes the electrical output of one or more strings of cells in the module to fall to zero, but not having the output of the entire module fall to zero.

At the end of September 2013, IKEA announced that solar panel packages for houses will be sold at 17 United Kingdom IKEA stores by the end of July 2014. The decision followed a successful pilot project at the Lakeside IKEA store, whereby one photovoltaic (PV) system was sold almost every day. The panels are manufactured by a Chinese company named Hanergy Holding Group Ltd.

Growth

Solar photovoltaics is growing rapidly, albeit from a small base, to a total global capacity of 139 gigawatts (GW) at the end of 2013. The total power output of the world's PV capacity in a calendar year is equal to some 160 billion kWh of electricity. This is sufficient to cover the annual power supply needs of 40 million households in the world, and represents 0.85 percent of worldwide electricity demand. More than 100 countries use solar PV. China, followed by Japan and the United States is now the fastest growing market, while Germany remains the world's largest producer, contributing almost 6 percent to its national electricity demands. Photovoltaics is now, after hydro and wind power, the third most important renewable energy source in terms of globally installed capacity.

The 2014 European Photovoltaic Industry Association (EPIA) report estimates global PV installations to grow 35-52 GW in 2014. China is predicted to take the lead from Germany and to become the world's largest producer of PV power in 2016. By 2018 the worldwide photovoltaic capacity is projected to have doubled (low scenario of 320 GW) or even tripled (high scenario of 430 GW) within five years. The EPIA also estimates that photovoltaics will meet 10 to 15 percent of Europe's energy demand in 2030.

The EPIA/Greenpeace Solar Generation Paradigm Shift Scenario (formerly called Advanced Scenario) from 2010 shows that by the year 2030, 1,845 GW of PV systems could be generating approximately 2,646 TWh/year of electricity around the world. Combined with energy use efficiency improvements, this would represent the electricity needs of more than 9 percent of the world's population. By 2050, over 20 percent of all electricity could be provided by photovoltaics.

Forecast

Michael Liebreich, from Bloomberg New Energy Finance, anticipates a tipping point for solar energy. The costs of power from wind and solar are already below those of conventional electricity generation in some parts of the world, as they have fallen sharply and will continue to do so. He also asserts, that the electrical grid has been greatly expanded worldwide, and is ready to receive and distribute electricity from renewable sources. In addition, worldwide electricity prices came under strong pressure from renewable energy sources, that are, in part, enthusiastically embraced by consumers. Deutsche Bank sees a "second gold rush" for the photovoltaic industry to come. Grid parity has already been reached in at least 19 markets by January 2014. Photovoltaics will prevail beyond feed-in tariffs, becoming more competitive as deployment increases and prices continue to fall.

In June 2014 Barclays downgraded bonds of U.S. utility companies. Barclays expects more competition by a growing self-consumption due to a combination of decentralized PV-systems and residential electricity storage. This could fundamentally change the utility's business model and transform the system over the next ten years, as prices for these systems are predicted to fall.

Economics

There have been major changes in the underlying costs, industry structure and market prices of solar photovoltaics technology, over the years, and gaining a coherent picture of the shifts occurring across the industry value chain globally is a challenge. This is due to: "the rapidity of cost and price changes, the complexity of the PV supply chain, which involves a large number of manufacturing processes, the balance of system (BOS) and installation costs associated with complete PV systems, the choice of different distribution channels, and differences between regional markets within which PV is being deployed". Further complexities result from the many different policy support initiatives that have been put in place to facilitate photovoltaics commercialisation in various countries.

The PV industry has seen dramatic drops in module prices since 2008. In late 2011, factory-gate prices for crystalline-silicon photovoltaic modules dropped below the $1.00/W mark. The $1.00/W installed cost, is often regarded in the PV industry as marking the achievement of grid parity for PV. Technological advancements, manufacturing process improvements, and industry re-structuring, mean that further price reductions are likely in coming years.

Financial incentives for photovoltaics, such as feed-in tariffs, have often been offered to electricity consumers to install and operate solar-electric generating systems. Government has sometimes also offered incentives in order to encourage the PV industry to achieve the economies of scale needed to compete where the cost of PV-generated electricity is above the cost from the existing grid. Such policies are implemented to promote national or territorial energy independence, high tech job creation and reduction of carbon dioxide emissions which cause global warming. Due to economies of scale solar panels get less costly as people use and buy more—as manufacturers increase production to meet demand, the cost and price is expected to drop in the years to come.

Solar cell efficiencies vary from 6% for amorphous silicon-based solar cells to 44.0% with multiple-junction concentrated photovoltaics. Solar cell energy conversion efficiencies for commercially available photovoltaics are around 14–22%. Concentrated photovoltaics (CPV) may reduce cost by concentrating up to 1,000 suns (through magnifying lens) onto a smaller sized photovoltaic cell. However, such concentrated solar power requires sophisticated heat sink designs, otherwise the photovoltaic cell overheats, which reduces its efficiency and life. To further exacerbate the concentrated cooling design, the heat sink must be passive, otherwise the power required for active cooling would reduce the overall efficiency and economy.

Crystalline silicon solar cell prices have fallen from $76.67/Watt in 1977 to an estimated $0.74/Watt in 2013. This is seen as evidence supporting Swanson's law, an observation similar to the famous Moore's Law that states that solar cell prices fall 20% for every doubling of industry capacity.

As of 2011, the price of PV modules per MW has fallen by 60% since the summer of 2008, according to Bloomberg New Energy Finance estimates, putting solar power for the first time on a competitive footing with the retail price of electricity in a number of sunny countries; an alternative and consistent price decline figure of 75% from 2007 to 2012 has also been published, though it is unclear whether these figures are specific to the United States or generally global. The levelised cost of electricity (LCOE) from PV is competitive with conventional electricity sources in an expanding list of geographic regions, particularly when the time of generation is included, as electricity is worth more during the day than at night. There has been fierce competition in the supply chain, and further improvements in the levelised cost of energy for solar lie ahead, posing a growing threat to the dominance of fossil fuel generation sources in the next few years. As time progresses, renewable energy technologies generally get cheaper, while fossil fuels generally get more expensive:

The less solar power costs, the more favourably it compares to conventional power, and the more attractive it becomes to utilities and energy users around the globe. Utility-scale solar power can now be delivered in California at prices well below $100/MWh ($0.10/kWh) less than most other peak generators, even those running on low-cost natural gas. Lower solar module costs also stimulate demand from consumer markets where the cost of solar compares very favourably to retail electric rates.

As of 2011, the cost of PV has fallen well below that of nuclear power and is set to fall further. The average retail price of solar cells as monitored by the Solarbuzz group fell from $3.50/watt to $2.43/ watt over the course of 2011. For large-scale installations, prices below $1.00/watt were achieved. A module price of 0.60 Euro/watt ($0.78/ watt) was published for a large scale 5-year deal in April 2012.

By the end of 2012, the "best in class" module price had dropped to $0.50/watt, and was expected to drop to $0.36/watt by 2017. In many locations, PV has reached grid parity, which is usually defined as PV production costs at or below retail electricity prices (though often still above the power station prices for coal or gas-fired generation without their distribution and other costs). However, in many countries there is still a need for more access to capital to develop PV projects. To solve this problem securitization has been proposed and used to accelerate development of solar photovoltaic projects. For example, SolarCity offered, the first U.S. asset-backed security in the solar industry in 2013.

Photovoltaic power is also generated during a time of day that is close to peak demand (precedes it) in electricity systems with high use of air conditioning. More generally, it is now evident that, given a carbon price of $50/ton, which would raise the price of coal-fired power by 5c/ kWh, solar PV will be cost-competitive in most locations. The declining price of PV has been reflected in rapidly growing installations, totaling about 23 GW in 2011. Although some consolidation is likely in 2012, due to support cuts in the large markets of Germany and Italy, strong growth seems likely to continue for the rest of the decade. Already, by one estimate, total investment in renewables for 2011 exceeded investment in carbon-based electricity generation.

In the case of self consumption payback time is calculated based on how much electricity is not brought from the grid. Additionally, using PV solar power to charge DC batteries, as used in Plug-in Hybrid Electric Vehicles and Electric Vehicles, leads to greater efficiencies. Traditionally, DC generated electricity from solar PV must be converted to AC for buildings, at an average 10% loss during the conversion. An additional efficiency loss occurs in the transition back to DC for battery driven devices and vehicles, and using various interest rates and energy price changes were calculated to find present values that range from $2,057.13 to $8,213.64 (analysis from 2009).

For example in Germany with electricity prices of 0.25 euro/KWh and Insolation of 900 KWh/KW one KWp will save 225 euro per year and with installation cost of 1700 euro/KWp means that the system will pay back in less than 7 years.

Applications

Power Stations: Many solar photovoltaic power stations have been built, mainly in Europe. As of July 2012, the largest photovoltaic (PV) power plants in the world are the Agua Caliente Solar Project (USA, 247 MW), Charanka Solar Park (India, 214 MW), Golmud Solar Park (China, 200 MW), Perovo Solar Park (Ukraine 100 MW), Sarnia Photovoltaic Power Plant (Canada, 97 MW), Brandenburg-Briest Solarpark (Germany 91 MW), Solarpark Finow Tower (Germany 84.7 MW), Montalto di Castro Photovoltaic Power Station (Italy, 84.2 MW), Eggebek Solar Park (Germany 83.6 MW), Senftenberg Solarpark (Germany 82 MW), Finsterwalde Solar Park (Germany, 80.7 MW), Okhotnykovo Solar Park (Ukraine, 80 MW), Lopburi Solar Farm (Thailand 73.16 MW), Rovigo Photovoltaic Power Plant (Italy, 72 MW), and the Lieberose Photovoltaic Park (Germany, 71.8 MW).

There are also many large plants under construction. The Desert Sunlight Solar Farm under construction in Riverside County, California and Topaz Solar Farm being built in San Luis Obispo County, California are both 550 MW solar parks that will use thin-film solar photovoltaic modules made by First Solar. The Blythe Solar Power Project is a 500 MW photovoltaic station under construction in Riverside County, California. The California Valley Solar Ranch (CVSR) is a 250 megawatt (MW) solar photovoltaic power plant, which is being built by SunPower in the Carrizo Plain, northeast of California Valley. The 230 MW Antelope Valley Solar Ranch is a First Solar photovoltaic project which is under construction in the Antelope Valley area of the Western Mojave Desert, and due to be completed in 2013. The Mesquite Solar project is a photovoltaic solar power plant being built in Arlington, Maricopa County, Arizona, owned by Sempra Generation. Phase 1 will have a nameplate capacity of 150 megawatts.

Many of these plants are integrated with agriculture and some use innovative tracking systems that follow the sun's daily path across the sky to generate more electricity than conventional fixed-mounted systems. There are no fuel costs or emissions during operation of the power stations.

Photovoltaic System

A photovoltaic system, also photovoltaic power system, solar PV system, PV system or casually solar array, is a power system designed to supply usable solar power by means of photovoltaics. It consists of an arrangement of several components, including solar panels to absorb and directly convert sunlight into electricity, a solar inverter to change the electrical current from DC to AC, as well as mounting, cabling and other electrical accessories to set-up a working system. It may also use a solar tracking system to improve the system's overall performance or include an integrated battery solution, as prices for storage devices are expected to decline. Strictly speaking, a solar array only encompasses the ensemble of solar panels, the visible part of the PV system, and does not include all the other hardware, often summarized as balance of system (BOS). Moreover, PV systems convert light directly into electricity and shouldn't be confused with other solar technologies, such as concentrated solar power (CSP) and solar thermal, used for both, heating and cooling.

Figure: *A photovoltaic system mounted in front of a bank of windows.*

PV systems range from small, roof-top mounted or building-integrated systems with capacities from a few to several tens of kilowatts, to large utility-scale power stations of hundreds of megawatts. Nowadays, most PV systems are connected to the electrical grid, while stand-alone or off-grid systems only account for a small portion of the market.

Operating silently and without any moving parts or environmental emissions, PV systems have developed into a mature technology that has been used for fifty years in specialised applications, and grid-connected systems have been operating for over twenty years. A roof-top system recoups the invested energy for its manufacturing and installation within 0.7 to 2 years and produces about 95 percent of net clean renewable energy over a 30-year service lifetime.

As new installations are growing exponentially, prices for PV systems have rapidly declined in recent years. However, they vary by market and the size of the system. In 2013 overall prices for installed rooftop systems were just below $5.00 per watt in the United States and Japan, while prices in the highly penetrated German market reached $2.00. Nowadays, solar panels account for less than half of the system's overall cost, leaving the rest to installation labour and to the PV system's remaining components.

Modern System

A photovoltaic system converts the sun's radiation into usable electricity. It comprises the solar array and the balance of system components. PV systems may be built in various configurations:

- Grid-connected optionally using a battery storage
- Off-grid without battery (array-direct)
- Off-grid with battery storage, optionally converting to AC

Besides these basic configurations, PV systems can be categorized by various aspects, such as, building-integrated vs. rack-mounted systems, residential vs. utility systems, distributed vs. centralized systems, roof-top vs. ground-mounted systems, tracking vs. fixed-tilt systems, new construction vs. retrofitted systems. Other distinctions may include, systems with microinverters vs. central inverter, systems using crystalline silicon vs. thin-film technology, and systems with modules from Chinese vs. European and US-manufacturers.

About 99 percent of all European and 90 percent of all US-PV systems are connected to the electrical grid, while in Australia and South Korea off-grid systems are somewhat more common. PV systems rarely use battery storage. This may change soon, as government incentives for distributed energy storage are being implemented and investments in storage solutions are gradually becoming economically viable for small systems. A solar array of a typical residential PV system is rack-mounted on the roof, rather than integrated into the roof or facade of the building, as this is significantly more expensive. Utility-scale solar power stations are ground-mounted, with fixed tilted solar panels rather than using expensive tracking devices. Crystalline silicon is the predominant material used in 90 percent of worldwide produced solar modules, while rival thin-film has lost market-share in recent years. About 70 percent of all solar cells and modules are produced in China and Taiwan, leaving only 5 percent to European and US-manufacturers. The installed capacity for both, small rooftop systems and large solar power station is growing rapidly and in equal parts. Photovoltaic systems currently contribute about 1 percent to worldwide electricity generation.

Driven by advances in technology and increases in manufacturing scale and sophistication, the cost of photovoltaics is declining continuously and the installed capacity of PV systems is growing exponentially. There are several million PV systems distributed all over the world, mostly in Europe, with 1.4 million systems in Germany alone– as well as North America with 440,000 systems in the United States, China and Japan, the world's fastest growing regions. The energy conversion efficiency of a conventional solar module increased from 15 to 20 percent over the last 10 years and a PV system recoups the energy needed for its manufacture in about 2 years. In exceptionally irradiated locations, or when thin-film technology is used, the so-called energy payback time decreases to one year or less. Net metering and financial incentives, such as preferential feed-in tariffs for solar-generated electricity, have also greatly supported installations of PV systems in many countries. The levelised cost of electricity (LCOE) from PV systems has become competitive with conventional electricity sources in an expanding list of geographic regions, as grid parity has been achieved in many different markets.

Grid-Connection

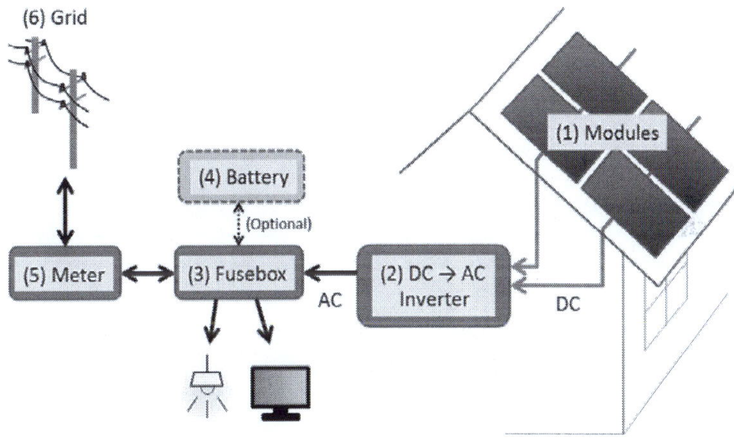

Figure: *Schematics of a typical residential PV system*

A grid connected system is connected to a larger independent grid (typically the public electricity grid) and feeds energy directly into the grid. This energy may be shared by a residential or commercial building before or after the revenue measurement point. The difference being whether the credited energy production is calculated independently of the customer's energy consumption (feed-in tariff) or only on the difference of energy (net metering). Grid connected systems vary in size from residential (2-10kWp) to solar power stations (up to 10s of MWp). This is a form of decentralized electricity generation. The feeding of electricity into the grid requires the transformation of DC into AC by a special, synchronising grid-tie inverter. In kW sized installations the DC side system voltage is as high as permitted (typically 1000V except US residential 600V) to limit ohmic losses. Most modules (72 crystalline silicon cells) generate 160W to 300W at 36 volts. It is sometimes necessary or desirable to connect the modules partially in parallel rather than all in series. One set of modules connected in series is known as a 'string'.

Scale of System

Photovoltaic systems are generally categorized into three distinct market segments: residential roof-top, commercial roof-top, and ground-mount utility-scale systems. Their capacities range from a few kilowatts to hundreds of megawats. A typical residential system is around 10 kilowatts and mounted on a sloped roof, while commercial systems may reach a megawatt-scale and are generally installed on low-slope or even flat roofs. Although roof-top mounted systems are small and display a higher cost per watt than large utility-scale installations, they account for the largest share in the market. There is, however, a growing trend towards bigger utility-scale power plants, especially in the "sunbelt" region of the planet.

Roof-Top System

A small PV system is capable of providing enough AC electricity to power a single home, or even an isolated device in the form of AC or DC electric. For example, military and civilian Earth observation satellites, street lights, construction and traffic signs, electric cars, solar-powered tents, and electric aircraft may contain integrated photovoltaic systems to provide a primary or auxiliary power source in the form of AC or DC power, depending on the design and power demands. In 2013, roof-top systems accounted for 60 percent of worldwide installations.

Building-Integrated System

In urban and suburban areas, photovoltaic arrays are commonly used on rooftops to supplement power use; often the building will have a connection to the power grid, in which case the energy produced by the PV array can be sold back to the utility in some sort of net metering agreement. Some utilities, such as Solvay Electric in Solvay, NY, use the rooftops of commercial customers and telephone poles to support their use of PV panels. Solar trees are arrays that, as the name implies, mimic the look of trees, provide shade, and at night can function as street lights.

Utility-Scale

Large utility-scale solar parks or farmss are power stations and capable of providing an energy supply to large numbers of consumers. Generated electricity is fed into the transmission grid powered by central generation plants (grid-connected or grid-tied plant), or combined with one, or many, domestic electricity generators to feed into a small electrical grid (hybrid plant). In rare cases generated electricity is stored or used directly by island/standalone plant. PV systems are generally designed in order to ensure the highest energy yield for a given investment. Some large photovoltaic power stations like Waldpolenz Solar Park and Topaz Solar Farm cover tens or hundreds of hectares and have power outputs up to hundreds of megawatts.

Performance

Uncertainties in revenue over time relate mostly to the evaluation of the solar resource and to the performance of the system itself. In the best of cases, uncertainties are typically 4% for year-to-year climate variability, 5% for solar resource estimation (in a horizontal plane), 3% for estimation of irradiation in the plane of the array, 3% for power rating of modules, 2% for losses due to dirt and soiling, 1.5% for losses due to snow, and 5% for other sources of error. Identifying and reacting to manageable losses is critical for revenue and O&M efficiency. Monitoring of array performance may be part of contractual agreements between the array owner, the builder, and the utility purchasing the energy produced. Recently, a method to create "synthetic days" using readily available weather data and verification using the Open Solar Outdoors Test Field make it possible to predict photovoltaic systems performance with high degrees of accuracy. This method can be used to then determine loss mechanisms on a local scale - such as those from snow or the effects of surface coatings (e.g. hydrophobic or hydrophilic) on soiling or snow losses. Access to the Internet has allowed a further improvement in energy monitoring and communication. Dedicated systems are available from a number of vendors. For solar PV system that use microinverters (panel-level DC to AC conversion), module power data is automatically provided. Some systems allow setting performance alerts that trigger phone/email/text warnings when limits are reached. These solutions provide data for the system owner and the installer. Installers are able to remotely monitor multiple installations, and see at-a-glance the status of their entire installed base.

Components

A photovoltaic system for residential, commercial, or industrial energy supply consists of the solar array and a number of components often summarized as the balance of system (BOS). The term originates from the fact that some BOS-components are balancing the power-generating subsystem of the solar array with the power-using side, the load. BOS-components include power-conditioning equipment and structures for mounting, typically one or more DC to AC power converters, also known as inverters, an energy storage device, a racking system that supports the solar array, electrical wiring and interconnections, and mounting for other components.

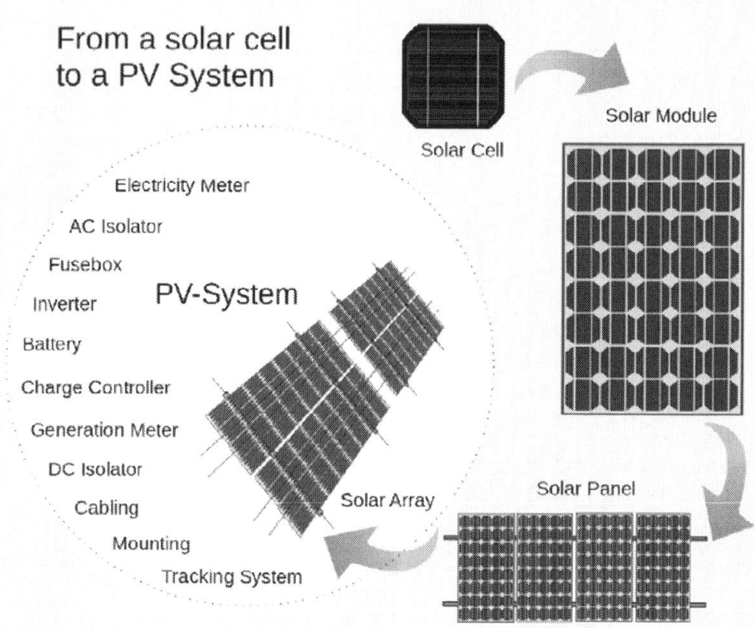

Figure: *Diagram of the possible components of a photovoltaic system.*

Optionally, a balance of system may include any or all of the following: renewable energy credit revenue-grade meter, maximum power point tracker (MPPT), battery system and charger, GPS solar tracker, energy management software, solar irradiance sensors, anemometer, or task-specific accessories designed to meet specialised requirements for a system owner. In addition, a CPV system requires optical lenses or mirrors and sometimes a cooling system.

The terms *"solar array"* and *"PV system"* are often used interchangeably, despite the fact that the solar array does not encompass the entire system. Moreover, *"solar panel"* is often used as a synonym for *"solar module"*, although a panel consists of a string of several modules. The term *"solar system"* is also an often used misnomer for a PV system.

Solar Array

Conventional c-Si solar cells, normally wired in series, are encapsulated in a solar module to protect them from the weather. The module consists of a tempered glass as cover, a soft and flexible encapsulant, a rear backsheet made of a weathering and fire-resistant material and an aluminum frame around the outer edge. Electrically connected and mounted on a supporting structure, solar modules build a string of modules, often called solar panel. A solar array consists of one or many such panels.

A *photovoltaic array* (or *solar array*) is a linked collection of solar panels. The power that one module can produce is seldom enough to meet requirements of a home or a business, so the modules are linked together to form an *array*.

Most PV arrays use an inverter to convert the DC power produced by the modules into alternating current that can power lights, motors, and other loads. The modules in a PV array are usually first connected in series to obtain the desired voltage; the individual strings are then connected in parallel to allow the system to produce more current. Solar panels are typically measured under STC (standard test conditions) or PTC (PVUSA test conditions), in watts. Typical panel ratings range from less than 100 watts to over 400 watts. The array rating consists of a summation of the panel ratings, in watts, kilowatts, or megawatts.

Insolation and Energy

Solar insolation is made up of direct, diffuse, and reflected radiation. The absorption factor of a PV cell is defined as the fraction of incident solar irradiance that is absorbed by the cell. At high noon on a cloudless day at the equator, the power of the sun is about 1 kW/m^2, on the Earth's surface, to a plane that is perpendicular to the sun's rays. As such, PV arrays can track the sun through each day to greatly enhance energy collection. However, tracking devices add cost, and require maintenance, so it is more common for PV arrays to have fixed mounts that tilt the array and face solar noon (approximately due south in the Northern Hemisphere or due north in the Southern Hemisphere). The tilt angle, from horizontal, can be varied for season, but if fixed, should be set to give optimal array output during the peak electrical demand portion of a typical year for a stand alone system. This optimal module tilt angle is not necessarily identical to the tilt angle for maximum annual array energy output.

The optimization of the a photovoltaic system for a specific environment can be complicated as issues of solar flux, soiling, and snow losses should be taken into effect. In addition, recent work has shown that spectral effects can play a role in optimal photovoltaic material selection. For example, the spectral albedo can play a significant role in output depending on the surface around the photovoltaic system and the type of solar cell material.

For the weather and latitudes of the United States and Europe, typical insolation ranges from 4 kWh/m^2/day in northern climes to 6.5 kWh/m^2/day in the sunniest regions. A photovoltaic installation in the southern latitudes of Europe or the United States may expect to produce 1 kWh/m^2/day. A typical 1 kW photovoltaic installation in Australia or the southern latitudes of Europe or United States, may produce 3.5-5 kWh per day, dependent on location, orientation, tilt, insolation and other factors. In the Sahara desert, with less cloud cover and a better solar angle, one could ideally obtain closer to 8.3 kWh/m^2/day provided the nearly ever present wind would not blow sand onto the units. The area of the Sahara desert is over 9 million km^2. 90,600 km^2, or about 1%, could generate as much electricity as all of the world's power plants combined.

Module and Efficiency

A typical "150 watt" solar module is about a square meter in size. Such a module may be expected to produce 0.75 kWh every day, on average, after taking into account the weather and the latitude, for an insolation of 5 sun hours/day. In the last 10 years, the efficiency of average commercial wafer-based crystalline silicon modules increased from about 12% to 16% and CdTe module efficiency increased from 9% to 13% during same period. Module output and life degraded by increased temperature. Allowing ambient air to flow over, and if possible behind, PV modules reduces this problem. Effective module lives are typically 25 years or more. The payback period for an investment in a PV solar installation varies greatly and is typically less useful than a calculation of return on investment. While it is typically calculated to be between 10 and 20 years, the financial payback period can be far shorter with incentives.

Due to the low voltage of an individual solar cell (typically ca. 0.5V), several cells are wired (see: Copper in photovoltaic power systems) in series in the manufacture of a "laminate". The laminate is assembled into a protective weatherproof enclosure, thus making a photovoltaic module or solar panel. Modules may then be strung together into a photovoltaic array.

In 2012, solar panels available for consumers can have an efficiency of up to about 17%, while commercially available panels can go as far as 27%. It has been recorded that a group from the The Fraunhofer Institute for Solar Energy Systems have created a cell that can reach 44.7% efficiency, which makes scientists' hopes of reaching the 50% efficiency threshold a lot more feasible.

Shading and Dirt

Photovoltaic cell electrical output is extremely sensitive to shading. The effects of this shading are well known. When even a small portion of a cell, module, or array is shaded, while the remainder is in sunlight, the output falls dramatically due to internal 'short-circuiting' (the electrons reversing course through the shaded portion of the p-n junction). If the current drawn from the series string of cells is no greater than the current that can be produced by the shaded cell, the current (and so power) developed by the string is limited. If enough voltage is available from the rest of the cells in a string, current will be forced through the cell by breaking down the junction in the shaded portion. This breakdown voltage in common cells is between 10 and 30 volts. Instead of adding to the power produced by the panel, the shaded cell absorbs power, turning it into heat. Since the reverse voltage of a shaded cell is much greater than the forward voltage of an illuminated cell, one shaded cell can absorb the power of many other cells in the string, disproportionately affecting panel output. For example, a shaded cell may drop 8 volts, instead of adding 0.5 volts, at a particular current level, thereby absorbing the power produced by 16 other cells. It is, thus important that a PV installation is not shaded by trees or other obstructions.

Several methods have been developed to determine shading losses from trees to PV systems over both large regions using LiDAR, but also at an individual system level using sketchup. Most modules have bypass diodes between each cell or string of cells that minimize the effects of shading and only lose the power of the shaded portion of the array. The main job of the bypass diode is to eliminate hot spots that form on cells that can cause further damage to the array, and cause fires. Sunlight can be absorbed by dust, snow, or other impurities at the surface of the module. This can reduce the light that strikes the cells. In general these losses aggregated over the year are small even for locations in Canada. Maintaining a clean module surface will increase output performance over the life of the module. Google found that cleaning the flat mounted solar panels after 15 months increased their output by almost 100%, but that the 5% tilted arrays were adequately cleaned by rainwater.

Cabling

Due to their outdoor usage, solar cables are specifically designed to be resistant against UV radiation and extremely high temperature fluctuations and are generally unaffected by the weather. A number of standards specify the usage of electrical wiring in PV systems, such as the IEC 60364 by the International Electrotechnical Commission, in section 712 *"Solar photovoltaic (PV) power supply systems"*, the British Standard BS 7671, incorporating regulations relating to microgeneration and photovoltaic systems, and the US UL4703 standard, in subject 4703 *"Photovoltaic Wire"*.

Mounting Systems

Modules are assembled into arrays on some kind of mounting system, which may be classified as ground mount, roof mount or pole mount. For solar parks a large rack is mounted on the ground, and the modules mounted on the rack. For buildings, many different racks have been devised for pitched roofs. For flat roofs, racks, bins and building integrated solutions are used. Solar panel racks mounted on top of poles can be stationary or moving, such as a light fixture or an antenna. Pole mounting raises what would otherwise be a ground mounted array above weed shadows and livestock, and may satisfy electrical code requirements regarding inaccessibility of exposed wiring. Pole mounted panels are open to more cooling air on their underside, which increases performance. A multiplicity of pole top racks can be formed into a parking carport or other shade structure. A rack which does not follow the sun from left to right may allow seasonal adjustment up or down.

Solar Trackers

A solar tracking system tilts a solar panel throughout the day. Depending on the type of tracking system, the panel is either aimed directly at the sun or the brightest area of a partly clouded sky. Trackers greatly enhance early morning and late afternoon performance, increasing the total amount of power produced by a system by about 20–25% for a single axis tracker and about 30% or more for a dual axis tracker, depending on latitude. Trackers are effective in regions that receive a large portion of sunlight directly. In diffuse light (i.e. under cloud or fog), tracking has little or no value. Because most concentrated photovoltaics systems are very sensitive to the sunlight's angle, tracking systems allow them to produce useful power for more than a brief period each day. Tracking systems improve performance for two main reasons. First, when a solar panel is perpendicular to the sunlight, it receives more light on its surface than if it were angled. Second, direct light is used more efficiently than angled light. Special Anti-reflective coatings can improve solar panel efficiency for direct and angled light, somewhat reducing the benefit of tracking.

Trackers and sensors to optimise the performance are often seen as optional, but tracking systems can increase viable output by up to 45%. PV arrays that approach or exceed one megawatt often use solar trackers. Accounting for clouds, and the fact that most of the world is not on the equator, and that the sun sets in the evening, the correct measure of solar power is insolation – the average number of kilowatt-hours per square meter per day. For the weather and latitudes of the United States and Europe, typical insolation ranges from 2.26 kWh/m^2/day in northern climes to 5.61 kWh/m^2/day in the sunniest regions.

For large systems, the energy gained by using tracking systems can outweigh the added complexity (trackers can increase efficiency by 30% or more). For very large systems, the added maintenance of tracking is a substantial detriment. Tracking is not required for flat panel and low-concentration photovoltaic systems. For high-concentration photovoltaic systems, dual axis tracking is a necessity.

Pricing trends affect the balance between adding more stationary solar panels versus having fewer panels that track. When solar panel prices drop, trackers become a less attractive option.

Solar Inverters

Systems designed to deliver alternating current (AC), such as grid-connected applications need an inverter to convert the direct current (DC) from the solar modules to AC. Grid connected inverters must supply AC electricity in sinusoidal form, synchronized to the grid frequency, limit feed in voltage to no higher than the grid voltage and disconnect from the grid if the grid voltage is turned off. Islanding inverters need only produce regulated voltages and frequencies in a sinusoidal waveshape as no synchronisation or co-ordination with grid supplies is required.

A solar inverter may connect to a string of solar panels. In some installations a solar micro-inverter is connected at each solar panel. For safety reasons a circuit breaker is provided both on the AC and DC side to enable maintenance. AC output may be connected through an electricity meter into the public grid. The number of modules in the system determines the total DC watts capable of being generated by the solar array; however, the inverter ultimately governs the amount of AC watts that can be distributed for consumption. For example: A PV system comprising 11 kilowatts DC (kWDC) worth of PV modules, paired with one 10-kilowatt AC (kWAC) inverter, will be limited by the maximum output of the inverter: 10 kW AC.

As of 2014, conversion efficiency for state-of-the-art converters reached more than 98 percent. While string inverters are used in residential to medium-sized commercial PV systems, central inverters cover the large commercial and utility-scale market. Market-share for central and string inverters are about 50 percent and 48 percent, respectively, leaving less than 2 percent to micro-inverters.

Inverter/Converter Market in 2014				
Type	***Power***	***Efficiency***[a]	***Market Share***[b]	***Remarks***
String inverter	up to 100 kW_p[c]	98%	50%	Cost[b] €0.15 per watt-peak. Easy to replace.
Central inverter	above 100 kW_p	98.5%	48%	€0.10 per watt-peak. High reliability. Often sold along with a service contract.
Micro-inverter	module power range	90%–95%	1.5%	€0.40 per watt-peak. Ease of replacement concerns.
DC/DC converter Power optimizer	module power range	98.8%	n.a.	€0.40 per watt-peak. Ease of replacement concerns. Inverter is still needed. About 0.75 GWP installed in 2013.
Source: *data by IHS 2014, remarks by Fraunhofer ISE 2014, from: Photovoltaics Report, updated as per 8 September 2014, p. 35, PDF* ***Notes:*** *[a]best efficiencies displayed, [b]market-share and cost per watt are estimated, [c]kW_p = kilowatt-peak*				

Maximum Power Point Tracking

Maximum power point tracking (MPPT) is a technique that grid connected inverters use to get the maximum possible power from the photovoltaic array. In order to do so, the Inverter's MPPT system digitally samples the solar array's ever changing power output and applies the proper resistance to find the optimal *maximum power point.*

Anti-Islanding

Anti-islanding is a protection mechanism, that immediately shuts down the inverter preventing it from generating AC-power, when the connection to the load no longer exists. This happens, for example, in the case of a blackout. Without this protection, the supply line would become an "island" with power surrounded by a "sea" of unpowered lines, as the solar array continues to deliver DC power during the power outage. Islanding is a hazard to utility workers, who may not realise that an AC-circuit is still powered, and it may prevent automatic re-connection of devices.

Charge Controller

PV systems with integrated battery solutions also need a charge controller, as the varying voltage and current from the solar array requires constant adjustment to prevent damage from overcharging. Basic charge controllers may simply turn the PV panels on and off, or may meter out pulses of energy as needed, a strategy called PWM or pulse-width modulation. More advanced charge controllers will incorporate MPPT logic into their battery charging algorithms. Charge controllers may also divert energy to some purpose other than battery charging. Rather than simply shut off the free PV energy when not needed, a user may choose to heat air or water once the battery is full.

Battery

Although still expensive, PV systems increasingly use rechargeable batteries to store a surplus to be later used at night. Batteries used for grid-storage also stabilize the electrical grid by leveling out peak loads, and play an important role in a smart grid, as they can charge during periods of low demand and feed their stored energy into the grid when demand is high.

Common battery technologies used in today's PV systems include, the valve regulated lead-acid battery– a modified version of the conventional lead–acid battery, nickel–cadmium and lithium-ion batteries. Compared to the other types, lead-acid batteries have a shorter lifetime and lower energy density. However, due to their high reliability, low self discharge as well as low investment and maintenance costs, they are currently the predominant technology used in small-scale, residential PV systems, as lithium-ion batteries are still being developed and about 3.5 times as expensive as lead-acid batteries. Furthermore, as storage devices for PV systems are used stationary, the lower energy and power density and therefore higher weight of lead-acid batteries are not as critical as, for example, in electric transportation

Other rechargeable batteries that are considered for distributed PV systems include, sodium–sulfur and vanadium redox batteries, two prominent types of a molten salt and a flow battery, respectively.

Solar Thermal Energy

Solar thermal energy (STE) is a form of energy and a technology for harnessing solar energy to generate thermal energy or electrical energy for use in industry, and in the residential and commercial sectors. The first installation of solar thermal energy equipment occurred in the Sahara Desert approximately in 1910 and was a steam engine without a kettle and fire but with a mirror system for sun light collection to heat water for the needed steam pressure. Because of the influence of World War I, liquid fuel was better developed and the Sahara project was abandoned, only to be reused several decades later.

Solar thermal collectors are classified by the United States Energy Information Administration as low-, medium-, or high-temperature collectors. Low-temperature collectors are flat plates generally used to heat swimming pools.

Medium-temperature collectors are also usually flat plates but are used for heating water or air for residential and commercial use. High-temperature collectors concentrate sunlight using mirrors or lenses and are generally used for fulfilling heat requirements up to 300 deg C / 20 bar pressure in industries, and for electric power production.

However, there is a term that used for both the applications. Concentrated Solar Thermal (CST) for fulfilling heat requirements in industries and Concentrated Solar Power (CSP) when the heat collected is used for power generation. CST and CSP are not replaceable in terms of application. The 377 MW Ivanpah Solar Power Facility is the largest solar power plant in the world, located in the Mojave Desert of California. Other large solar thermal plants include the SEGS installation (354 MW), also in the Mojave, as well as the Solnova Solar Power Station (150 MW), the Andasol solar power station (150 MW), and Extresol Solar Power Station (100 MW), all in Spain.

Low-Temperature Solar Heating and Cooling Systems

Systems for utilising low-temperature solar thermal energy include means for heat collection; usually heat storage, either short-term or interseasonal; and distribution within a structure or a district heating network. In some cases more than one of these functions is inherent to a single feature of the system (e.g. some kinds of solar collectors also store heat). Some systems are passive, others are active (requiring other external energy to function).

Heating is the most obvious application, but solar cooling can be achieved for a building or district cooling network by using a heat-driven absorption or adsorption chiller (heat pump). There is a productive coincidence that the greater the driving heat from insulation, the greater the cooling output. In 1878, Auguste Mouchout pioneered solar cooling by making ice using a solar steam engine attached to a refrigeration device.

In the United States, heating, ventilation, and air conditioning (HVAC) systems account for over 25% (4.75 EJ) of the energy used in commercial buildings and nearly half (10.1 EJ) of the energy used in residential buildings. Solar heating, cooling, and ventilation technologies can be used to offset a portion of this energy.

In Europe, since the mid-1990s about 125 large solar-thermal district heating plants have been constructed, each with over 500 m2 (5400 ft2) of solar collectors. The largest are about 10,000 m2, with capacities of 7 MW-thermal and solar heat costs around 4 Eurocents/ kWh without subsidies. 40 of them have nominal capacities of 1 MW-thermal or more. The Solar District Heating program (SDH) has participation from 14 European Nations and the European Commission, and is working toward technical and market development, and holds annual conferences.

Low-Temperature Collectors

Glazed Solar Collectors are designed primarily for space heating and they recirculate building air through a solar air panel where the air is heated and then directed back into the building. These solar space heating systems require at least two penetrations into the building and only perform when the air in the solar collector is warmer than the building room temperature. Most glazed collectors are used in the residential sector.

Unglazed Solar Collectors are primarily used to pre-heat make-up ventilation air in commercial, industrial and institutional buildings with a high ventilation load. They turn building walls or sections of walls into low cost, high performance, unglazed solar collectors. Also called, "transpired solar panels", they employ a painted perforated metal solar heat absorber that also serves as the exterior wall surface of the building. Heat conducts from the absorber surface to the thermal boundary layer of air 1 mm thick on the outside of the absorber and to air that passes behind the absorber. The boundary layer of air is drawn into a nearby perforation before the heat can escape by convection to the outside air. The heated air is then drawn from behind the absorber plate into the building's ventilation system.

A Trombe wall is a passive solar heating and ventilation system consisting of an air channel sandwiched between a window and a sun-facing thermal mass. During the ventilation cycle, sunlight stores heat in the thermal mass and warms the air channel causing circulation through vents at the top and bottom of the wall. During the heating cycle the Trombe wall radiates stored heat.

Solar roof ponds are unique solar heating and cooling systems developed by Harold Hay in the 1960s. A basic system consists of a roof-mounted water bladder with a movable insulating cover. This system can control heat exchange between interior and exterior environments by covering and uncovering the bladder between night and day. When heating is a concern the bladder is uncovered during the day allowing sunlight to warm the water bladder and store heat for evening use. When cooling is a concern the covered bladder draws heat from the building's interior during the day and is uncovered at night to radiate heat to the cooler atmosphere. The Skytherm house in Atascadero, California uses a prototype roof pond for heating and cooling.

Solar space heating with solar air heat collectors is more popular in the USA and Canada than heating with solar liquid collectors since most buildings already have a ventilation system for heating and cooling. The two main types of solar air panels are glazed and unglazed.

Of the 21,000,000 square feet (2,000,000 m^2) of solar thermal collectors produced in the United States in 2007, 16,000,000 square feet (1,500,000 m^2) were of the low-temperature variety. Low-temperature collectors are generally installed to heat swimming pools, although they can also be used for space heating. Collectors can use air or water as the medium to transfer the heat to their destination.

Heat Storage in Low-Temperature Solar Thermal Systems

Interseasonal storage. Solar heat (or heat from other sources) can be effectively stored between opposing seasons aquifers, underground geological strata, large specially constructed pits, and large tanks that are insulated and covered with earth.

Short-term storage. Thermal mass materials store solar energy during the day and release this energy during cooler periods. Common thermal mass materials include stone, concrete, and water. The proportion and placement of thermal mass should consider several factors such as climate, daylighting, and shading conditions. When properly incorporated, thermal mass can passively maintain comfortable temperatures while reducing energy consumption.

Solar-Driven Cooling

Worldwide, by 2011 there were about 750 cooling systems with solar-driven heat pumps, and annual market growth was 40 to 70% over the prior seven years. It is a niche market because the economics are challenging, with the annual number of cooling hours a limiting factor. Respectively, the annual cooling hours are roughly 1000 in the Mediterranean, 2500 in Southeast Asia, and only 50 to 200 in Central Europe. However, system construction costs dropped about 50% between 2007 and 2011. The International Energy Agency (IEA) Solar

Heating and Cooling program (IEA-SHC) task groups working on further development of the technologies involved. The International Energy Agency (IEA) Solar Heating and Cooling program (IEA-SHC) task groups working on further development of the technologies involved.

Solar Heat-Driven Ventilation

A solar chimney (or thermal chimney) is a passive solar ventilation system composed of a hollow thermal mass connecting the interior and exterior of a building. As the chimney warms, the air inside is heated causing an updraft that pulls air through the building. These systems have been in use since Roman times and remain common in the Middle East.

Process Heat

Solar process heating systems are designed to provide large quantities of hot water or space heating for nonresidential buildings.

Evaporation ponds are shallow ponds that concentrate dissolved solids through evaporation. The use of evaporation ponds to obtain salt from sea water is one of the oldest applications of solar energy. Modern uses include concentrating brine solutions used in leach mining and removing dissolved solids from waste streams. Altogether, evaporation ponds represent one of the largest commercial applications of solar energy in use today.

Unglazed transpired collectors (UTC) are perforated sun-facing walls used for preheating ventilation air. UTCs can raise the incoming air temperature up to 22 °C and deliver outlet temperatures of 45-60 °C. The short payback period of transpired collectors (3 to 12 years) make them a more cost-effective alternative to glazed collection systems. As of 2009, over 1500 systems with a combined collector area of 300,000 m^2 had been installed worldwide. Representatives include an 860 m^2 collector in Costa Rica used for drying coffee beans and a 1300 m^2 collector in Coimbatore, India used for drying marigolds.

A food processing facility in Modesto, California uses parabolic troughs to produce steam used in the manufacturing process. The 5,000 m^2 collector area is expected to provide 15 TJ per year.

Medium-Temperature Collectors

These collectors could be used to produce approximately 50% and more of the hot water needed for residential and commercial use in the United States. In the United States, a typical system costs $4000–

$6000 retail ($1400 to $2200 wholesale for the materials) and 30% of the system qualifies for a federal tax credit + additional state credit exists in about half of the states. Labour for a simple open loop system in southern climates can take 3–5 hours for the installation and 4–6 hours in Northern areas. Northern system require more collector area and more complex plumbing to protect the collector from freezing. With this incentive, the payback time for a typical household is four to nine years, depending on the state.

Similar subsidies exist in parts of Europe. A crew of one solar plumber and two assistants with minimal training can install a system per day. Thermosiphon installation have negligible maintenance costs (costs rise if antifreeze and mains power are used for circulation) and in the US reduces a households' operating costs by $6 per person per month.

Solar water heating can reduce CO_2 emissions of a family of four by 1 ton/year (if replacing natural gas) or 3 ton/year (if replacing electricity). Medium-temperature installations can use any of several designs: common designs are pressurized glycol, drain back, batch systems and newer low pressure freeze tolerant systems using polymer pipes containing water with photovoltaic pumping.

European and International standards are being reviewed to accommodate innovations in design and operation of medium temperature collectors. Operational innovations include "permanently wetted collector" operation. This innovation reduces or even eliminates the occurrence of no-flow high temperature stresses called stagnation which would otherwise reduce the life expectancy of collectors.

Solar Drying

Solar thermal energy can be useful for drying wood for construction and wood fuels such as wood chips for combustion. Solar is also used for food products such as fruits, grains, and fish. Crop drying by solar means is environmentally friendly as well as cost effective while improving the quality.

The less money it takes to make a product, the less it can be sold for, pleasing both the buyers and the sellers. Technologies in solar drying include ultra low cost pumped transpired plate air collectors based on black fabrics. Solar thermal energy is helpful in the process of drying products such as wood chips and other forms of biomass by raising the temperature while allowing air to pass through and get rid of the moisture.

Cooking

Solar cookers use sunlight for cooking, drying and pasteurization. Solar cooking offsets fuel costs, reduces demand for fuel or firewood, and improves air quality by reducing or removing a source of smoke.

The simplest type of solar cooker is the box cooker first built by Horace de Saussure in 1767. A basic box cooker consists of an insulated container with a transparent lid. These cookers can be used effectively with partially overcast skies and will typically reach temperatures of 50–100 °C.

Concentrating solar cookers use reflectors to concentrate solar energy onto a cooking container. The most common reflector geometries are flat plate, disc and parabolic trough type. These designs cook faster and at higher temperatures (up to 350 °C) but require direct light to function properly.

The Solar Kitchen in Auroville, India uses a unique concentrating technology known as the solar bowl. Contrary to conventional tracking reflector/fixed receiver systems, the solar bowl uses a fixed spherical reflector with a receiver which tracks the focus of light as the Sun moves across the sky. The solar bowl's receiver reaches temperature of 150 °C that is used to produce steam that helps cook 2,000 daily meals.

Many other solar kitchens in India use another unique concentrating technology known as the Scheffler reflector. This technology was first developed by Wolfgang Scheffler in 1986. A Scheffler reflector is a parabolic dish that uses single axis tracking to follow the Sun's daily course. These reflectors have a flexible reflective surface that is able to change its curvature to adjust to seasonal variations in the incident angle of sunlight. Scheffler reflectors have the advantage of having a fixed focal point which improves the ease of cooking and are able to reach temperatures of 450-650 °C. Built in 1999 by the Brahma Kumaris, the world's largest Scheffler reflector system in Abu Road, Rajasthan India is capable of cooking up to 35,000 meals a day. By early 2008, over 2000 large cookers of the Scheffler design had been built worldwide.

Distillation

Solar stills can be used to make drinking water in areas where clean water is not common. Solar distillation is necessary in these situations to provide people with purified water. Solar energy heats up the water in the still. The water then evaporates and condenses on the bottom of the covering glass.

High-Temperature Collectors

Where temperatures below about 95 °C are sufficient, as for space heating, flat-plate collectors of the nonconcentrating type are generally used. Because of the relatively high heat losses through the glazing, flat plate collectors will not reach temperatures much above 200 °C even when the heat transfer fluid is stagnant. Such temperatures are too low for efficient conversion to electricity.

The efficiency of heat engines increases with the temperature of the heat source. To achieve this in solar thermal energy plants, solar radiation is concentrated by mirrors or lenses to obtain higher temperatures – a technique called Concentrated Solar Power (CSP). The practical effect of high efficiencies is to reduce the plant's collector size and total land use per unit power generated, reducing the environmental impacts of a power plant as well as its expense.

As the temperature increases, different forms of conversion become practical. Up to 600 °C, steam turbines, standard technology, have an efficiency up to 41%. Above 600 °C, gas turbines can be more efficient. Higher temperatures are problematic because different materials and techniques are needed. One proposal for very high temperatures is to use liquid fluoride salts operating between 700 °C to 800 °C, using multi-stage turbine systems to achieve 50% or more thermal efficiencies. The higher operating temperatures permit the plant to use higher-temperature dry heat exchangers for its thermal exhaust, reducing the plant's water use – critical in the deserts where large solar plants are practical. High temperatures also make heat storage more efficient, because more watt-hours are stored per unit of fluid.

Commercial concentrating solar thermal power (CSP) plants were first developed in the 1980s. The world's largest solar thermal power plants are now the 370 MW Ivanpah Solar Power Facility, commissioned in 2014, and the 354 MW SEGS CSP installation, both located in the Mojave Desert of California, where several other solar projects have been realised as well. With the exception of the Shams solar power station, built in 2013 near Abu Dhabi, the United Arab Emirates, all other 100 MW or larger CSP plants are either located in the United States or in Spain.

The principal advantage of CSP is the ability to efficiently add thermal storage, allowing the dispatching of electricity over up to a

24-hour period. Since peak electricity demand typically occurs at about 5 pm, many CSP power plants use 3 to 5 hours of thermal storage. With current technology, storage of heat is much cheaper and more efficient than storage of electricity. In this way, the CSP plant can produce electricity day and night. If the CSP site has predictable solar radiation, then the CSP plant becomes a reliable power plant. Reliability can further be improved by installing a back-up combustion system. The back-up system can use most of the CSP plant, which decreases the cost of the back-up system.

CSP facilities utilise high electrical conductivity materials, such as copper, in field power cables, grounding networks, and motors for tracking and pumping fluids, as well as in the main generator and high voltage transformers.

With reliability, unused desert, no pollution, and no fuel costs, the obstacles for large deployment for CSP are cost, aesthetics, land use and similar factors for the necessary connecting high tension lines. Although only a small percentage of the desert is necessary to meet global electricity demand, still a large area must be covered with mirrors or lenses to obtain a significant amount of energy. An important way to decrease cost is the use of a simple design.

When considering land use impacts associated with the exploration and extraction through to transportation and conversion of fossil fuels, which are used for most of our electrical power, utility-scale solar power compares as one of the most land-efficient energy resources available:

The federal government has dedicated nearly 2,000 times more acreage to oil and gas leases than to solar development. In 2010 the Bureau of Land Management approved nine large-scale solar projects, with a total generating capacity of 3,682 megawatts, representing approximately 40,000 acres. In contrast, in 2010, the Bureau of Land Management processed more than 5,200 applications gas and oil leases, and issued 1,308 leases, for a total of 3.2 million acres. Currently, 38.2 million acres of onshore public lands and an additional 36.9 million acres of offshore exploration in the Gulf of Mexico are under lease for oil and gas development, exploration and production.

System Designs

During the day the sun has different positions. For low concentration systems (and low temperatures) tracking can be avoided (or limited to a few positions per year) if nonimaging optics are used. For higher concentrations, however, if the mirrors or lenses do not move, then the focus of the mirrors or lenses changes (but also in these cases nonimaging optics provides the widest acceptance angles for a given concentration). Therefore it seems unavoidable that there needs to be a tracking system that follows the position of the sun (for solar photovoltaic a solar tracker is only optional). The tracking system increases the cost and complexity. With this in mind, different designs can be distinguished in how they concentrate the light and track the position of the sun.

Parabolic Trough Designs

Parabolic trough power plants use a curved, mirrored trough which reflects the direct solar radiation onto a glass tube containing a fluid (also called a receiver, absorber or collector) running the length of the trough, positioned at the focal point of the reflectors. The trough is parabolic along one axis and linear in the orthogonal axis.

For change of the daily position of the sun perpendicular to the receiver, the trough tilts east to west so that the direct radiation remains focused on the receiver. However, seasonal changes in the in angle of sunlight parallel to the trough does not require adjustment of the mirrors, since the light is simply concentrated elsewhere on the receiver. Thus the trough design does not require tracking on a second axis. The receiver may be enclosed in a glass vacuum chamber. The vacuum significantly reduces convective heat loss.

A fluid (also called heat transfer fluid) passes through the receiver and becomes very hot. Common fluids are synthetic oil, molten salt and pressurized steam. The fluid containing the heat is transported to a heat engine where about a third of the heat is converted to electricity.

Full-scale parabolic trough systems consist of many such troughs laid out in parallel over a large area of land. Since 1985 a solar thermal system using this principle has been in full operation in California in the United States. It is called the Solar Energy Generating Systems (SEGS) system. Other CSP designs lack this kind of long experience and therefore it can currently be said that the parabolic trough design is the most thoroughly proven CSP technology.

The SEGS is a collection of nine plants with a total capacity of 354 MW and has been the world's largest solar power plant, both thermal and non-thermal, for many years. A newer plant is Nevada Solar One plant with a capacity of 64 MW. The 150 MW Andasol solar power stations are in Spain with each site having a capacity of 50 MW. Note however, that those plants have heat storage which requires a larger field of solar collectors relative to the size of the steam turbine-generator to store heat and send heat to the steam turbine at the same time. Heat storage enables better utilisation of the steam turbine. With day and some night-time operation of the steam-turbine Andasol 1 at 50 MW peak capacity produces more energy than Nevada Solar One at 64 MW peak capacity, due to the former plant's thermal energy storage system and larger solar field. The 280MW Solana Generating Station came online in Arizona in 2013 with 6 hours of power storage. Hassi R'Mel integrated solar combined cycle power station in Algeria and Martin Next Generation Solar Energy Centree both use parabolic troughs in a combined cycle with natural gas.

Power Tower Designs

Power towers (also known as 'central tower' power plants or 'heliostat' power plants) capture and focus the sun's thermal energy with thousands of tracking mirrors (called heliostats) in roughly a two square mile field. A tower resides in the centree of the heliostat field. The heliostats focus concentrated sunlight on a receiver which sits on top of the tower. Within the receiver the concentrated sunlight heats molten salt to over 1,000 °F (538 °C). The heated molten salt then flows into a thermal storage tank where it is stored, maintaining 98% thermal efficiency, and eventually pumped to a steam generator. The steam drives a standard turbine to generate electricity. This process, also known as the "Rankine cycle" is similar to a standard coal-fired power plant, except it is fuelled by clean and free solar energy.

The advantage of this design above the parabolic trough design is the higher temperature. Thermal energy at higher temperatures can be converted to electricity more efficiently and can be more cheaply stored for later use. Furthermore, there is less need to flatten the ground area. In principle a power tower can be built on the side of a hill. Mirrors can be flat and plumbing is concentrated in the tower. The disadvantage is that each mirror must have its own dual-axis control, while in the parabolic trough design single axis tracking can be shared for a large array of mirrors.

A cost/performance comparison between power tower and parabolic trough concentrators was made by the NREL which estimated that by 2020 electricity could be produced from power towers for 5.47 ¢/kWh and for 6.21 ¢/kWh from parabolic troughs. The capacity factor for power towers was estimated to be 72.9% and 56.2% for parabolic troughs. There is some hope that the development of cheap, durable, mass producible heliostat power plant components could bring this cost down.

The first commercial tower power plant was PS10 in Spain with a capacity of 11 MW, completed in 2007. Since then a number of plants have been proposed, several have been built on a number of countries (Spain, Germany, U.S., Turkey, China, India) but several proposed plants were cancelled as PV solar prices plummeted. A solar power tower is expected to come online in South Africa in 2014. Ivanpah Solar Power Facility in California generates 392 MW of electricity from three towers, making it the largest solar power tower plant when it came online in late 2013.

Dish Designs

CSP-Stirling is known to have the highest efficiency of all solar technologies around 30% compared to solar PV approximately 15%, and is predicted to be able to produce the cheapest energy among all renewable energy sources in high scale production and hot areas, semi deserts etc. A dish Stirling system uses a large, reflective, parabolic dish (similar in shape to satellite television dish). It focuses all the sunlight that strikes the dish up onto a single point above the dish, where a receiver captures the heat and transforms it into a useful form. Typically the dish is coupled with a Stirling engine in a Dish-Stirling System, but also sometimes a steam engine is used. These create rotational kinetic energy that can be converted to electricity using an electric generator.

In 2005 Southern California Edison announced an agreement to purchase solar powered Stirling engines from Stirling Energy Systems over a twenty-year period and in quantities (20,000 units) sufficient to generate 500 megawatts of electricity. In January 2010, Stirling Energy Systems and Tessera Solar commissioned the first demonstration 1.5-megawatt power plant ("Maricopa Solar") using Stirling technology in Peoria, Arizona. At the beginning of 2011 Stirling Energy's development arm, Tessera Solar, sold off its two large projects, the 709 MW Imperial project and the 850 MW Calico project to AES Solar and K.Road, respectively, and in the fall of 2011 Stirling Energy Systems applied for Chapter 7 bankruptcy due to competition from low cost photovoltaics. In 2012 the Maricopa plant was bought and dismantled by United Sun Systems.

Fresnel Technologies

A linear Fresnel reflector power plant uses a series of long, narrow, shallow-curvature (or even flat) mirrors to focus light onto one or more linear receivers positioned above the mirrors. On top of the receiver a small parabolic mirror can be attached for further focusing the light. These systems aim to offer lower overall costs by sharing a receiver between several mirrors (as compared with trough and dish concepts), while still using the simple line-focus geometry with one axis for tracking. This is similar to the trough design (and different from central towers and dishes with dual-axis). The receiver is stationary and so fluid couplings are not required (as in troughs and dishes). The mirrors also do not need to support the receiver, so they are structurally simpler. When suitable aiming strategies are used (mirrors aimed at different receivers at different times of day), this can allow a denser packing of mirrors on available land area.

Rival single axis tracking technologies include the relatively new linear Fresnel reflector (LFR) and compact-LFR (CLFR) technologies. The LFR differs from that of the parabolic trough in that the absorber is fixed in space above the mirror field. Also, the reflector is composed of many low row segments, which focus collectively on an elevated long tower receiver running parallel to the reflector rotational axis.

Prototypes of Fresnel lens concentrators have been produced for the collection of thermal energy by International Automated Systems. No full-scale thermal systems using Fresnel lenses are known to be in operation, although products incorporating Fresnel lenses in conjunction with photovoltaic cells are already available.

MicroCSP

MicroCSP is used for community-sized power plants (1 MW to 50 MW), for industrial, agricultural and manufacturing 'process heat' applications, and when large amounts of hot water are needed, such as resort swimming pools, water parks, large laundry facilities, sterilization, distillation and other such uses.

Enclosed Parabolic Trough

The enclosed parabolic trough solar thermal system encapsulates the components within an off-the-shelf greenhouse type of glasshouse. The glasshouse protects the components from the elements that can negatively impact system reliability and efficiency. This protection importantly includes nightly glass-roof washing with optimized water-efficient off-the-shelf automated washing systems. Lightweight curved solar-reflecting mirrors are suspended from the ceiling of the glasshouse by wires. A single-axis tracking system positions the mirrors to retrieve the optimal amount of sunlight. The mirrors concentrate the sunlight and focus it on a network of stationary steel pipes, also suspended from the glasshouse structure. Water is pumped through the pipes and boiled to generate steam when intense sun radiation is applied. The steam is available for process heat. Sheltering the mirrors from the wind allows them to achieve higher temperature rates and prevents dust from building up on the mirrors as a result from exposure to humidity.

Heat Collection and Exchange

More energy is contained in higher frequency light based upon the formula of $E = hnu$, where h is the Planck constant and υ is frequency. Metal collectors down convert higher frequency light by producing a series of Compton shifts into an abundance of lower frequency light. Glass or ceramic coatings with high transmission in the visible and UV and effective absorption in the IR (heat blocking) trap metal absorbed low frequency light from radiation loss. Convection insulation prevents mechanical losses transferred through gas. Once collected as heat, thermos containment efficiency improves significantly with increased size. Unlike Photovoltaic technologies that often degrade under concentrated light, Solar Thermal depends upon light concentration that requires a clear sky to reach suitable temperatures.

Heat in a solar thermal system is guided by five basic principles: heat gain; heat transfer; heat storage; heat transport; and heat insulation. Here, heat is the measure of the amount of thermal energy an object contains and is determined by the temperature, mass and specific heat of the object.

Solar thermal power plants use heat exchangers that are designed for constant working conditions, to provide heat exchange. Copper heat exchangers are important in solar thermal heating and cooling systems because of copper's high thermal conductivity, resistance to atmospheric and water corrosion, sealing and joining by soldering, and mechanical strength. Copper is used both in receivers and in primary circuits (pipes and heat exchangers for water tanks) of solar thermal water systems.

Heat gain is the heat accumulated from the sun in the system. Solar thermal heat is trapped using the greenhouse effect; the greenhouse effect in this case is the ability of a reflective surface to transmit short wave radiation and reflect long wave radiation. Heat and infrared radiation (IR) are produced when short wave radiation light hits the absorber plate, which is then trapped inside the collector. Fluid, usually water, in the absorber tubes collect the trapped heat and transfer it to a heat storage vault.

Heat is transferred either by conduction or convection. When water is heated, kinetic energy is transferred by conduction to water molecules throughout the medium. These molecules spread their thermal energy by conduction and occupy more space than the cold slow moving molecules above them.

The distribution of energy from the rising hot water to the sinking cold water contributes to the convection process. Heat is transferred from the absorber plates of the collector in the fluid by conduction. The collector fluid is circulated through the carrier pipes to the heat transfer vault. Inside the vault, heat is transferred throughout the medium through convection.

Heat storage enables solar thermal plants to produce electricity during hours without sunlight. Heat is transferred to a thermal storage medium in an insulated reservoir during hours with sunlight, and is withdrawn for power generation during hours lacking sunlight. Thermal storage mediums will be discussed in a heat storage section. Rate of heat transfer is related to the conductive and convection medium as well as the temperature differences. Bodies with large temperature differences transfer heat faster than bodies with lower temperature differences.

Heat transport refers to the activity in which heat from a solar collector is transported to the heat storage vault. Heat insulation is vital in both heat transport tubing as well as the storage vault. It prevents heat loss, which in turn relates to energy loss, or decrease in the efficiency of the system.

Heat Storage for Space Heating

A collection of mature technologies called seasonal thermal energy storage (STES) is capable of storing heat for months at a time, so solar heat collected primarily in Summer can be used for all-year heating. Solar-supplied STES technology has been advanced primarily in Denmark, Germany, and Canada, and applications include individual buildings and district heating networks. Drake Landing Solar Community in Alberta, Canada has a small district system and in 2012 achieved a world record of providing 97% of the a community's all-year space heating needs from the sun. STES thermal storage mediums include deep aquifers; native rock surrounding clusters of small-diameter, heat exchanger equipped boreholes; large, shallow, lined pits that are filled with gravel and top-insulated; and large, insulated and buried surface water tanks.

Heat Storage to Stabilize Solar-Electric Power Generation

Heat storage allows a solar thermal plant to produce electricity at night and on overcast days. This allows the use of solar power for baseload generation as well as peak power generation, with the potential of displacing both coal- and natural gas-fired power plants. Additionally, the utilisation of the generator is higher which reduces cost.

Heat is transferred to a thermal storage medium in an insulated reservoir during the day, and withdrawn for power generation at night. Thermal storage media include pressurized steam, concrete, a variety of phase change materials, and molten salts such as calcium, sodium and potassium nitrate.

Steam Accumulator

The PS10 solar power tower stores heat in tanks as pressurized steam at 50 bar and 285 °C. The steam condenses and flashes back to steam, when pressure is lowered. Storage is for one hour. It is suggested that longer storage is possible, but that has not been proven yet in an existing power plant.

Molten Salt Storage

A variety of fluids have been tested to transport the sun's heat, including water, air, oil, and sodium, but Rockwell International selected molten salt as best. Molten salt is used in solar power tower systems because it is liquid at atmospheric pressure, provides a low-cost medium to store thermal energy, its operating temperatures are compatible with today's steam turbines, and it is non-flammable and nontoxic. Molten salt is used in the chemical and metals industries to transport heat, so industry has experience with it.

The first commercial molten salt mixture was a common form of saltpeter, 60% sodium nitrate and 40% potassium nitrate. Saltpeter melts at 220 °C (430 °F) and is kept liquid at 290 °C (550 °F) in an insulated storage tank. Calcium nitrate can reduce the melting point to 131 °C, permitting more energy to be extracted before the salt freezes. There are now several technical calcium nitrate grades stable at more than 500 °C.

This solar power system can generate power in cloudy weather or at night using the heat in the tank of hot salt. The tanks are insulated, able to store heat for a week. Tanks that power a 100-megawatt turbine for four hours would be about 9 m (30 ft) tall and 24 m (80 ft) in diameter.

The Andasol power plant in Spain is the first commercial solar thermal power plant using molten salt for heat storage and night-time generation. It came on line March 2009. On July 4, 2011, a company in Spain celebrated an historic moment for the solar industry: Torresol's 19.9 MW concentrating solar power plant became the first ever to generate uninterrupted electricity for 24 hours straight, using a molten salt heat storage.

Phase-Change Materials for Storage

Phase Change Material (PCMs) offer an alternative solution in energy storage. Using a similar heat transfer infrastructure, PCMs have the potential of providing a more efficient means of storage. PCMs can be either organic or inorganic materials. Advantages of organic PCMs include no corrosives, low or no undercooling, and chemical and thermal stability. Disadvantages include low phase-change enthalpy, low thermal conductivity, and flammability. Inorganics are advantageous with greater phase-change enthalpy, but exhibit disadvantages with undercooling, corrosion, phase separation, and lack of thermal stability. The greater phase-change enthalpy in inorganic PCMs make hydrate salts a strong candidate in the solar energy storage field.

Use of Water

A design which requires water for condensation or cooling may conflict with location of solar thermal plants in desert areas with good solar radiation but limited water resources. The conflict is illustrated by plans of Solar Millennium, a German company, to build a plant in the Amargosa Valley of Nevada which would require 20% of the water available in the area. Some other projected plants by the same and other companies in the Mojave Desert of California may also be affected by difficulty in obtaining adequate and appropriate water rights. California water law currently prohibits use of potable water for cooling.

Other designs require less water. The proposed Ivanpah Solar Power Facility in south-eastern California will conserve scarce desert water by using air-cooling to convert the steam back into water. Compared to conventional wet-cooling, this results in a 90% reduction in water usage at the cost of some loss of efficiency. The water is then returned to the boiler in a closed process which is environmentally friendly.

Conversion Rates from Solar Energy to Electrical Energy

Of all of these technologies the solar dish/Stirling engine has the highest energy efficiency. A single solar dish-Stirling engine installed at Sandia National Laboratories National Solar Thermal Test Facility (NSTTF) produces as much as 25 kW of electricity, with a conversion efficiency of 31.25%.

Solar parabolic trough plants have been built with efficiencies of about 20%. Fresnel reflectors have an efficiency that is slightly lower (but this is compensated by the denser packing).

The gross conversion efficiencies (taking into account that the solar dishes or troughs occupy only a fraction of the total area of the power plant) are determined by net generating capacity over the solar energy that falls on the total area of the solar plant. The 500-megawatt (MW) SCE/SES plant would extract about 2.75% of the radiation (1 kW/m^2) that falls on its 4,500 acres (18.2 km^2). For the 50 MW AndaSol Power Plant that is being built in Spain (total area of 1,300×1,500 m = 1.95 km^2) gross conversion efficiency comes out at 2.6%.

Furthermore, efficiency does not directly relate to cost: on calculating total cost, both efficiency and the cost of construction and maintenance should be taken into account.

Chapter 4

Energy from Fission and Fusion

Binding Energy

Binding energy is the energy required to disassemble a whole system into separate parts. A bound system typically has a lower potential energy than the sum of its constituent parts — this is what keeps the system together. Often this means that energy is released upon the creation of a bound state. This definition corresponds to a *positive* binding energy. (This definition also often causes confusion. For example: A prominent term in chemistry is the 'free energy of binding', which is the difference between the bound and unbound states and thus *negative*).

General Idea

In general, binding energy represents the mechanical work that must be done against the forces which hold an object together, disassembling the object into component parts separated by sufficient distance that further separation requires negligible additional work.

At the atomic level the atomic binding energy of the atom derives from electromagnetic interaction and is the energy required to disassemble an atom into free electrons and a nucleus. Electron binding energy is a measure of the energy required to free electrons from their atomic orbits. This is more commonly known as ionization energy.

At the nuclear level, binding energy is also equal to the energy liberated when a nucleus is created from other nucleons or nuclei. This nuclear binding energy (binding energy of nucleons into a nuclide) is derived from the nuclear force (residual strong interaction) and is the energy required to disassemble a nucleus into the same number of free, unbound neutrons and protons it is composed of, so that the nucleons are far/distant enough from each other so that the nuclear force can no longer cause the particles to interact.

In astrophysics, gravitational binding energy of a celestial body is the energy required to expand the material to infinity. This quantity is not to be confused with the gravitational potential energy, which is the energy required to separate two bodies, such as a celestial body and a satellite, to infinite distance, keeping each intact (the latter energy is lower).

In bound systems, if the binding energy is removed from the system, it must be subtracted from the mass of the unbound system, simply because this energy *has* mass. Thus, if energy is removed (or emitted) from the system at the time it is bound, the loss of energy from the system will also result in the loss of the mass of the energy, from the system. System mass is not conserved in this process because the system is "open" (i.e., is not an isolated system to mass or energy input or loss) during the binding process.

Mass-Energy Relation

Classically a bound system is at a lower energy level than its unbound constituents, and its mass must be less than the total mass of its unbound constituents. For systems with low binding energies, this "lost" mass after binding may be fractionally small. For systems with high binding energies, however, the missing mass may be an easily measurable fraction.

Since all forms of energy exhibit rest mass within systems at "rest" (that is, in systems which have no net momentum), the question of where the missing mass of the binding energy goes, is of interest. The answer is that this mass is lost from a system which is not closed. It transforms to heat, light, higher energy states of the nucleus/atom or other forms of energy, but these types of energy also have mass, and it is necessary that they be removed from the system before its mass may decrease. The "mass deficit" from binding energy is therefore removed mass that corresponds with removed energy, according to Einstein's equation $E = mc^2$. Once the system cools to normal temperatures and returns to ground states in terms of energy levels, there is less mass remaining in the system than there was when it first combined and was at high energy. Mass measurements are almost always made at low temperatures with systems in ground states, and this difference between the mass of a system and the sum of the masses of its isolated parts is called a mass deficit. Thus, if binding energy mass is transformed into heat, the system must be cooled (the heat removed) before the mass-deficit appears in the cooled system. In that case, the removed heat represents exactly the mass "deficit", and the heat itself retains the mass which was lost (from the point of view of the initial system). This mass appears in any other system which absorbs the heat and gains thermal energy.

As an illustration, consider two objects attracting each other in space through their gravitational field. The attraction force accelerates the objects and they gain some speed toward each other converting the potential (gravity) energy into kinetic (movement) energy. When either the particles 1) pass through each other without interaction or 2) elastically repel during the collision, the gained kinetic energy (related to speed), starts to revert into potential form driving the collided particles apart. The decelerating particles will return to the initial distance and beyond into infinity or stop and repeat the collision (oscillation takes place). This shows that the system, which loses no energy, does not combine (bind) into a solid object, parts of which oscillate at short distances.

Therefore, in order to bind the particles, the kinetic energy gained due to the attraction must be dissipated (by resistive force). Complex objects in collision ordinarily undergo inelastic collision, transforming some kinetic energy into internal energy (heat content, which is atomic movement), which is further radiated in the form of photons—the light and heat. Once the energy to escape the gravity is dissipated in the collision, the parts will oscillate at closer, possibly atomic, distance, thus looking like one solid object.

This lost energy, necessary to overcome the potential barrier in order to separate the objects, is the binding energy. If this binding energy were retained in the system as heat, its mass would not decrease. However, binding energy lost from the system (as heat radiation) would itself have mass, and directly represents the "mass deficit" of the cold, bound system.

Closely analogous considerations apply in chemical and nuclear considerations. Exothermic chemical reactions in closed systems do not change mass, but become less massive once the heat of reaction is removed, though this mass change is much too small to measure with standard equipment. In nuclear reactions, however, the fraction of mass that may be removed as light or heat, i.e., binding energy, is often a much larger fraction of the system mass. It may thus be measured directly as a mass difference between rest masses of reactants and (cooled) products. This is because nuclear forces are comparatively stronger than the Coulombic forces associated with the interactions between electrons and protons, that generate heat in chemistry.

Mass Change

Mass change (decrease) in bound systems, particularly atomic nuclei, has also been termed *mass defect, mass deficit,* or mass *packing fraction.*

The difference between the unbound system calculated mass and experimentally measured mass of nucleus (mass change) is denoted by Δm. It can be calculated as follows:

Mass change = (unbound system calculated mass) - (measured mass of system) i.e., (sum of masses of protons and neutrons) - (measured mass of nucleus)

After nuclear reactions that result in an excited nucleus, the energy that must be radiated or otherwise removed as binding energy for a single nucleus may be in the form of electromagnetic waves, such as gamma radiation, or it may appear in the kinetic energy of an ejected particle, such as an electron, in internal conversion decay.

Also, energy of excitation of nucleus can be partly emitted as the rest mass of one or more a particle, such as the emitted particles of beta decay. Again, however, no mass deficit can in theory appear until this radiation or this energy has been emitted, and is no longer part of the system.

When nucleons bind together to form a nucleus, they must lose a small amount of mass, i.e., there is mass change, in order to stay bound. This mass change must be released as various types of photon or other particle energy as above, according to the relation $E = mc^2$. Thus, after binding energy has been removed, binding energy = mass change $\times c^2$. This energy is a measure of the forces that hold the nucleons together, and it represents energy which must be supplied again from the environment, if the nucleus were to be broken up into individual nucleons.

The energy given off during either nuclear fusion or nuclear fission is the difference between the binding energies of the "fuel", i.e., the initial nuclide(s), and the fission or fusion products. In practice, this energy may also be calculated from the substantial mass differences between the fuel and products, which uses previous measurement of the atomic masses of known nuclides, which always have the same mass for each species. This mass difference appears once evolved heat and radiation have been removed, which is a given requirement for measuring the (rest) masses of the (non-excited) nuclides involved in such calculations.

In 2005, Rainville et al. published a direct test of the energy-equivalence of mass lost in the binding-energy of a neutron to atoms of particular isotopes of silicon and sulfur, by comparing the new mass-change to the energy of the emitted gamma ray associated with the neutron capture. The binding mass-loss agreed with the gamma ray energy to a precision of ±0.00004 %, the most accurate test of $E=mc^2$ to date.

Excess Mass

It is observed experimentally that the mass of the nucleus is smaller than the number of nucleons each counted with a mass of 1 a.m.u.. This difference is called mass excess.

The difference between the actual mass of the nucleus measured in atomic mass units and the number of nucleons is called mass excess i.e.

Mass excess = M - A = Excess-energy / c^2

with : M equals the actual mass of the nucleus, in u.

and : A equals the mass number.

This mass excess is a practical value calculated from experimentally measured nucleon masses and stored in nuclear databases. For middle-weight nuclides this value is negative in contrast to the mass change which is never negative for any nuclide.

If a certain number of neutrons and protons are bought together to form a nucleus of certain range and mass, an energy Eb will be released in this process. The energy Eb is called the Nuclear Binding Energy.==Nuclear binding energy==

Nuclear Reactor

A nuclear reactor is a device to initiate and control a sustained nuclear chain reaction. Nuclear reactors are used at nuclear power plants for electricity generation and in propulsion of ships. Heat from nuclear fission is passed to a working fluid (water or gas), which runs through turbines. These either drive a ship's propellers or turn electrical generators. Nuclear generated steam in principle can be used for industrial process heat or for district heating. Some reactors are used to produce isotopes for medical and industrial use, or for production of plutonium for weapons. Some are run only for research. Today there are about 450 nuclear power reactors that are used to generate electricity in about 30 countries around the world.

Figure: *Core of CROCUS, a small nuclear reactor used for research at the EPFL in Switzerland*

Echanism

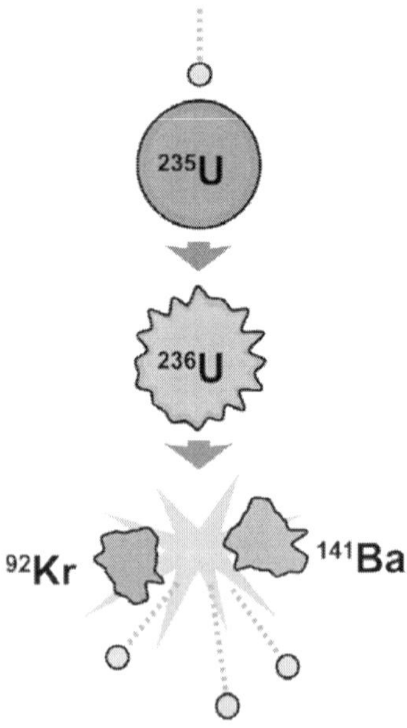

Figure: *An induced nuclear fission event. A neutron is absorbed by the nucleus of a uranium-235 atom, which in turn splits into fast-moving lighter elements (fission products) and free neutrons. Though both reactors and nuclear weapons rely on nuclear chain-reactions, the rate of reactions in a reactor occurs much more slowly than in a bomb.*

Just as conventional power-stations generate electricity by harnessing the thermal energy released from burning fossil fuels, nuclear reactors convert the thermal energy released from nuclear fission.

Fission

When a large fissile atomic nucleus such as uranium-235 or plutonium-239 absorbs a neutron, it may undergo nuclear fission. The heavy nucleus splits into two or more lighter nuclei, (the fission products), releasing kinetic energy, gamma radiation, and free neutrons. A portion of these neutrons may later be absorbed by other fissile atoms and trigger further fission events, which release more neutrons, and so on. This is known as a nuclear chain reaction.

To control such a nuclear chain reaction, neutron poisons and neutron moderators can change the portion of neutrons that will go on to cause more fission. Nuclear reactors generally have automatic and manual systems to shut the fission reaction down if monitoring detects unsafe conditions.

Commonly-used moderators include regular (light) water (in 74.8% of the world's reactors), solid graphite (20% of reactors) and heavy water (5% of reactors). Some experimental types of reactor have used beryllium, and hydrocarbons have been suggested as another possibility.

Heat Generation

The reactor core generates heat in a number of ways:

- The kinetic energy of fission products is converted to thermal energy when these nuclei collide with nearby atoms.
- The reactor absorbs some of the gamma rays produced during fission and converts their energy into heat.
- Heat is produced by the radioactive decay of fission products and materials that have been activated by neutron absorption. This decay heat-source will remain for some time even after the reactor is shut down.

A kilogram of uranium-235 (U-235) converted via nuclear processes releases approximately three million times more energy than a kilogram of coal burned conventionally (7.2×10^{13} joules per kilogram of uranium-235 versus 2.4×10^{7} joules per kilogram of coal).

Cooling

A nuclear reactor coolant — usually water but sometimes a gas or a liquid metal (like liquid sodium) or molten salt — is circulated past the reactor core to absorb the heat that it generates. The heat is carried away from the reactor and is then used to generate steam. Most reactor systems employ a cooling system that is physically separated from the water that will be boiled to produce pressurized steam for the turbines, like the pressurized water reactor. However, in some reactors the water for the steam turbines is boiled directly by the reactor core; for example the boiling water reactor.

Reactivity Control

The power output of the reactor is adjusted by controlling how many neutrons are able to create more fissions.

Control rods that are made of a neutron poison are used to absorb neutrons. Absorbing more neutrons in a control rod means that there are fewer neutrons available to cause fission, so pushing the control rod deeper into the reactor will reduce its power output, and extracting the control rod will increase it.

At the first level of control in all nuclear reactors, a process of delayed neutron emission by a number of neutron-rich fission isotopes is an important physical process. These delayed neutrons account for about 0.65% of the total neutrons produced in fission, with the remainder (termed "prompt neutrons") released immediately upon fission. The fission products which produce delayed neutrons have half lives for their decay by neutron emission that range from milliseconds to as long as several minutes. Keeping the reactor in the zone of chain-reactivity where delayed neutrons are *necessary* to achieve a critical mass state, allows time for mechanical devices or human operators to have time to control a chain reaction in "real time"; otherwise the time between achievement of criticality and nuclear meltdown as a result of an exponential power surge from the normal nuclear chain reaction, would be too short to allow for intervention.

In some reactors, the coolant also acts as a neutron moderator. A moderator increases the power of the reactor by causing the fast neutrons that are released from fission to lose energy and become thermal neutrons. Thermal neutrons are more likely than fast neutrons to cause fission. If the coolant is a moderator, then temperature changes can affect the density of the coolant/moderator and therefore change power output. A higher temperature coolant would be less dense, and therefore a less effective moderator.

In other reactors the coolant acts as a poison by absorbing neutrons in the same way that the control rods do. In these reactors power output can be increased by heating the coolant, which makes it a less dense poison. Nuclear reactors generally have automatic and manual systems to scram the reactor in an emergency shut down. These systems insert large amounts of poison (often boron in the form of boric acid) into the reactor to shut the fission reaction down if unsafe conditions are detected or anticipated.

Most types of reactors are sensitive to a process variously known as xenon poisoning, or the iodine pit. Xenon-135 produced in the fission process acts as a "neutron poison" that absorbs neutrons and therefore tends to shut the reactor down. Xenon-135 accumulation can be controlled by keeping power levels high enough to destroy it as fast as it is produced. Fission also produces iodine-135, which in turn decays (with a half-life of under seven hours) to new xenon-135. When the reactor is shut down, iodine-135 continues to decay to xenon-135, making restarting the reactor more difficult for a day or two. This temporary state is the "iodine pit." If the reactor has sufficient extra reactivity capacity, it can be restarted. As the extra xenon-135 is transmuted to xenon-136 which is not a neutron poison, within a few hours the reactor experiences a "xenon burnoff (power) transient". Control rods must be further inserted to replace the neutron absorption of the lost xenon-135. Failure to properly follow such a procedure was a key step in the Chernobyl disaster.

Reactors used in nuclear marine propulsion (especially nuclear submarines) often cannot be run at continuous power around the clock in the same way that land-based power reactors are normally run, and in addition often need to have a very long core life without refuelling. For this reason many designs use highly enriched uranium but incorporate burnable neutron poison directly into the fuel rods. This allows the reactor to be constructed with a high excess of fissionable material, which is nevertheless made relatively more safe early in the reactor's fuel burn-cycle by the presence of the neutron-absorbing material which is later replaced by naturally produced long-lived neutron poisons (far longer-lived than xenon-135) which gradually accumulate over the fuel load's operating life.

Electrical Power Generation

The energy released in the fission process generates heat, some of which can be converted into usable energy. A common method of harnessing this thermal energy is to use it to boil water to produce pressurized steam which will then drive a steam turbine that turns an alternator and generates electricity.

Early Reactors

The neutron was discovered in 1932. The concept of a nuclear chain reaction brought about by nuclear reactions mediated by neutrons was first realised shortly thereafter, by Hungarian scientist Leó Szilárd, in 1933. He filed a patent for his idea of a simple nuclear reactor the following year while working at the Admiralty in London. However, Szilárd's idea did not incorporate the idea of nuclear fission as a neutron source, since that process was not yet discovered. Szilárd's ideas for nuclear reactors using neutron-mediated nuclear chain reactions in light elements proved unworkable.

Inspiration for a new type of reactor using uranium came from the discovery by Lise Meitner, Fritz Strassmann and Otto Hahn in 1938 that bombardment of uranium with neutrons (provided by an alpha-on-beryllium fusion reaction, a "neutron howitzer") produced a barium residue, which they reasoned was created by the fissioning of the uranium nuclei. Subsequent studies in early 1939 (one of them by Szilárd and Fermi) revealed that several neutrons were also released during the fissioning, making available the opportunity for the nuclear chain reaction that Szilárd had envisioned six years previously.

On 2 August 1939 Albert Einstein signed a letter to President Franklin D. Roosevelt (written by Szilárd) suggesting that the discovery of uranium's fission could lead to the development of "extremely powerful bombs of a new type", giving impetus to the study of reactors and fission. Szilárd and Einstein knew each other well and had worked together years previously, but Einstein had never thought about this possibility for nuclear energy until Szilard reported it to him, at the beginning of his quest to produce the Einstein-Szilárd letter to alert the U.S. government.

Shortly after, Hitler's Germany invaded Poland in 1939, starting World War II in Europe. The U.S. was not yet officially at war, but in October, when the Einstein-Szilárd letter was delivered to him, Roosevelt commented that the purpose of doing the research was to make sure "the Nazis don't blow us up." The U.S. nuclear project followed, although with some delay as there remained skepticism (some of it from Fermi) and also little action from the small number of officials in the government who were initially charged with moving the project forward.

The following year the U.S. Government received the Frisch–Peierls memorandum from the UK, which stated that the amount of uranium needed for a chain reaction was far lower than had previously been thought. The memorandum was a product of the MAUD Committee, which was working on the UK atomic bomb project, known as Tube Alloys, later to be subsumed within the Manhattan Project.

Eventually, the first artificial nuclear reactor, Chicago Pile-1, was constructed at the University of Chicago, by a team led by Enrico Fermi, in late 1942. By this time, the program had been pressured for a year by U.S. entry into the war. The Chicago Pile achieved criticality on 2 December 1942 at 3:25 PM. The reactor support structure was made of wood, which supported a pile (hence the name) of graphite blocks, embedded in which was natural uranium-oxide 'pseudospheres' or 'briquettes'.

Soon after the Chicago Pile, the U.S. military developed a number of nuclear reactors for the Manhattan Project starting in 1943. The primary purpose for the largest reactors (located at the Hanford Site in Washington state), was the mass production of plutonium for nuclear weapons. Fermi and Szilard applied for a patent on reactors on 19 December 1944. Its issuance was delayed for 10 years because of wartime secrecy.

"World's first nuclear power plant" is the claim made by signs at the site of the EBR-I, which is now a museum near Arco, Idaho. Originally called "Chicago Pile-5", it was carried out under the direction of Walter Zinn for Argonne National Laboratory. This experimental LMFBR operated by the U.S. Atomic Energy Commission produced 0.8 kW in a test on 20 December 1951 and 100 kW (electrical) the following day, having a design output of 200 kW (electrical).

Besides the military uses of nuclear reactors, there were political reasons to pursue civilian use of atomic energy. U.S. President Dwight Eisenhower made his famous Atoms for Peace speech to the UN General Assembly on 8 December 1953. This diplomacy led to the dissemination of reactor technology to U.S. institutions and worldwide.

The first nuclear power plant built for civil purposes was the AM-1 Obninsk Nuclear Power Plant, launched on 27 June 1954 in the Soviet Union. It produced around 5 MW (electrical).

After World War II, the U.S. military sought other uses for nuclear reactor technology. Research by the Army and the Air Force never came to fruition; however, the U.S. Navy succeeded when they steamed the USS *Nautilus* (SSN-571) on nuclear power 17 January 1955.

The first commercial nuclear power station, Calder Hall in Sellafield, England was opened in 1956 with an initial capacity of 50 MW (later 200 MW).

The first portable nuclear reactor "Alco PM-2A" used to generate electrical power (2 MW) for Camp Century from 1960.

Components

The key components common to most types of nuclear power plants are:

- Nuclear fuel
- Nuclear reactor core
- Neutron moderator
- Neutron poison
- Neutron howitzer (provides steady source of neutrons to re-initiate reaction following shutdown)
- Coolant (often the Neutron Moderator and the Coolant are the same, usually both purified water)
- Control rods
- Reactor vessel
- Boiler feedwater pump
- Steam generators (not in BWRs)
- Steam turbine
- Electrical generator
- Condenser
- Cooling tower (not always required)
- Radwaste System (a section of the plant handling radioactive waste)
- Refuelling Floor
- Spent fuel pool
- Nuclear safety systems
    - Reactor Protective System (RPS)
    - Emergency Diesel Generators
    - Emergency Core Cooling Systems (ECCS)
    - Standby Liquid Control System (emergency boron injection, in BWRs only)
- Essential service water system (ESWS)
- Containment building
- Control room
- Emergency Operations Facility

- Nuclear training facility (usually contains a Control Room simulator)

Figure: *The control room of NC State's Pulstar Nuclear Reactor.*

Reactor Types

Classifications: Nuclear Reactors are classified by several methods; a brief outline of these classification methods is provided.

Classification by Type of Nuclear Reaction:

- Nuclear fission. All commercial power reactors are based on nuclear fission. They generally use uranium and its product plutonium as nuclear fuel, though a thorium fuel cycle is also possible. Fission reactors can be divided roughly into two classes, depending on the energy of the neutrons that sustain the fission chain reaction:
    - o Thermal reactors (the most common type of nuclear reactor) use slowed or thermal neutrons to keep up the fission of their fuel. Almost all current reactors are of this type. These contain neutron moderator materials that slow neutrons until their neutron temperature is *thermalized*, that is, until

their kinetic energy approaches the average kinetic energy of the surrounding particles. Thermal neutrons have a far higher cross-section (probability) of fissioning the fissile nuclei uranium-235, plutonium-239, and plutonium-241, and a relatively lower probability of neutron capture by uranium-238 (U-238) compared to the faster neutrons that originally result from fission, allowing use of low-enriched uranium or even natural uranium fuel. The moderator is often also the coolant, usually water under high pressure to increase the boiling point. These are surrounded by a reactor vessel, instrumentation to monitor and control the reactor, radiation shielding, and a containment building.

Figure: *NC State's PULSTAR Reactor is a 1 MW pool-type research reactor with 4% enriched, pin-type fuel consisting of UO2 pellets in zircaloy cladding.*

- o Fast neutron reactors use fast neutrons to cause fission in their fuel. They do not have a neutron moderator, and use less-moderating coolants. Maintaining a chain reaction requires the fuel to be more highly enriched in fissile material (about 20% or more) due to the relatively lower probability of fission versus capture by U-238. Fast reactors have the potential to produce less transuranic waste because all actinides are fissionable with fast neutrons, but they are more difficult to build and more expensive to operate. Overall, fast reactors are less common than thermal reactors in most applications. Some early power stations were fast reactors, as are some Russian naval propulsion units. Construction of prototypes is continuing.

- Nuclear fusion. Fusion power is an experimental technology, generally with hydrogen as fuel. While not suitable for power production, Farnsworth-Hirsch fusors are used to produce neutron radiation.

Classification by Moderator Material: Used by thermal reactors:

- Graphite-moderated reactors
- Water moderated reactors
  - o Heavy-water reactors (Used in Canada.)
  - o Light-water-moderated reactors (LWRs). Light-water reactors (the most common type of thermal reactor) use ordinary water to moderate and cool the reactors. When at operating temperature, if the temperature of the water increases, its density drops, and fewer neutrons passing through it are slowed enough to trigger further reactions. That negative feedback stabilizes the reaction rate. Graphite and heavy-water reactors tend to be more thoroughly thermalised than light water reactors. Due to the extra thermalization, these types can use natural uranium/ unenriched fuel.
- Light-element-moderated reactors. These reactors are moderated by lithium or beryllium.
  - o Molten salt reactors (MSRs) are moderated by a light elements such as lithium or beryllium, which are constituents of the coolant/fuel matrix salts LiF and BeF_2.
  - o Liquid metal cooled reactors, such as one whose coolant is a mixture of Lead and Bismuth, may use BeO as a moderator.
- Organically moderated reactors (OMR) use biphenyl and terphenyl as moderator and coolant.

Classification by Coolant

- Water cooled reactor. There are 104 operating reactors in the United States. Of these, 69 are pressurized water reactors (PWR), and 35 are boiling water reactors (BWR).

- Pressurized water reactor (PWR) Pressurized water reactors constitute the large majority of all Western nuclear power plants.
  - A primary characteristic of PWRs is a pressurizer, a specialised pressure vessel. Most commercial PWRs and naval reactors use pressurizers. During normal operation, a pressurizer is partially filled with water, and a steam bubble is maintained above it by heating the water with submerged heaters. During normal operation, the pressurizer is connected to the primary reactor pressure vessel (RPV) and the pressurizer "bubble" provides an expansion space for changes in water volume in the reactor. This arrangement also provides a means of pressure control for the reactor by increasing or decreasing the steam pressure in the pressurizer using the pressurizer heaters.

Figure: *Treatment of the interior part of a VVER-1000 reactor frame on Atommash.*

  - Pressurised heavy water reactors are a subset of pressurized water reactors, sharing the use of a pressurized, isolated heat transport loop, but using heavy water as coolant and moderator for the greater neutron economies it offers.

- o Boiling water reactor (BWR)
  - BWRs are characterized by boiling water around the fuel rods in the lower portion of a primary reactor pressure vessel. A boiling water reactor uses U, enriched as uranium dioxide, as its fuel. The fuel is assembled into rods housed in a steel vessel that is submerged in water. The nuclear fission causes the water to boil, generating steam. This steam flows through pipes into turbines. The turbines are driven by the steam, and this process generates electricity. During normal operation, pressure is controlled by the amount of steam flowing from the reactor pressure vessel to the turbine.
- o Pool-type reactor

- Liquid metal cooled reactor. Since water is a moderator, it cannot be used as a coolant in a fast reactor. Liquid metal coolants have included sodium, NaK, lead, lead-bismuth eutectic, and in early reactors, mercury.
  - o Sodium-cooled fast reactor
  - o Lead-cooled fast reactor
- Gas cooled reactors are cooled by a circulating inert gas, often helium in high-temperature designs, while carbon dioxide has been used in past British and French nuclear power plants. Nitrogen has also been used. Utilisation of the heat varies, depending on the reactor. Some reactors run hot enough that the gas can directly power a gas turbine. Older designs usually run the gas through a heat exchanger to make steam for a steam turbine.
- Molten salt reactors (MSRs) are cooled by circulating a molten salt, typically a eutectic mixture of fluoride salts, such as FLiBe. In a typical MSR, the coolant is also used as a matrix in which the fissile material is dissolved.

Classification by Generation

- Generation I reactor (early prototypes, research reactors, non-commercial power producing reactors)
- Generation II reactor (most current nuclear power plants 1965–1996)
- Generation III reactor (evolutionary improvements of existing designs 1996-now)

- Generation IV reactor (technologies still under development unknown start date, possibly 2030)

The "Gen IV"-term was dubbed by the United States Department of Energy (DOE) for developing new plant types in 2000. In 2003, the French Commissariat à l'Énergie Atomique (CEA) was the first to refer to Gen II types in Nucleonics Week; first mentioning of Gen III was also in 2000 in conjunction with the launch of the Generation IV International Forum (GIF) plans.

Classification by Phase of Fuel

- Solid fuelled
- Fluid fuelled
  - o Aqueous homogeneous reactor
  - o Molten salt reactor
- Gas fuelled (theoretical)

Classification by Use

- Electricity
  - o Nuclear power plants including small modular reactors
- Propulsion
  - o Nuclear marine propulsion
  - o Various proposed forms of rocket propulsion
- Other uses of heat
  - o Desalination
  - o Heat for domestic and industrial heating
  - o Hydrogen production for use in a hydrogen economy
- Production reactors for transmutation of elements
  - o Breeder reactors are capable of producing more fissile material than they consume during the fission chain reaction (by converting fertile U-238 to Pu-239, or Th-232 to U-233). Thus, a uranium breeder reactor, once running, can be re-fuelled with natural or even depleted uranium, and a thorium breeder reactor can be re-fuelled with thorium; however, an initial stock of fissile material is required.
  - o Creating various radioactive isotopes, such as americium for use in smoke detectors, and cobalt-60, molybdenum-99 and others, used for imaging and medical treatment.
  - o Production of materials for nuclear weapons such as weapons-grade plutonium

- Providing a source of neutron radiation (for example with the pulsed Godiva device) and positron radiation (e.g. neutron activation analysis and potassium-argon dating)
- Research reactor: Typically reactors used for research and training, materials testing, or the production of radioisotopes for medicine and industry. These are much smaller than power reactors or those propelling ships, and many are on university campuses. There are about 280 such reactors operating, in 56 countries. Some operate with high-enriched uranium fuel, and international efforts are underway to substitute low-enriched fuel.

Current Technologies

Pressurized water reactors (PWR): These reactors use a pressure vessel to contain the nuclear fuel, control rods, moderator, and coolant. They are cooled and moderated by high-pressure liquid water. The hot radioactive water that leaves the pressure vessel is looped through a steam generator, which in turn heats a secondary (non-radioactive) loop of water to steam that can run turbines. They are the majority of current reactors. This is a thermal neutron reactor design, the newest of which are the VVER-1200, Advanced Pressurized Water Reactor and the European Pressurized Reactor. United States Naval reactors are of this type.

Figure: *Diablo Canyon — a PWR*

Boiling water reactors (BWR): A BWR is like a PWR without the steam generator. A boiling water reactor is cooled and moderated by water like a PWR, but at a lower pressure, which allows the water to boil inside the pressure vessel producing the steam that runs the turbines. Unlike a PWR, there is no primary and secondary loop. The thermal efficiency of these reactors can be higher, and they can be simpler, and even potentially more stable and safe. This is a thermal neutron reactor design, the newest of which are the Advanced Boiling Water Reactor and the Economic Simplified Boiling Water Reactor.

Figure: *Laguna Verde nuclear power plant — a BWR*

Pressurized Heavy Water Reactor (PHWR)

Figure: *The CANDU Qinshan Nuclear Power Plant*

A Canadian design (known as CANDU), these reactors are heavy-water-cooled and -moderated pressurized-water reactors. Instead of using a single large pressure vessel as in a PWR, the fuel is contained in hundreds of pressure tubes. These reactors are fuelled with natural uranium and are thermal neutron reactor designs. PHWRs can be refuelled while at full power, which makes them very efficient in their use of uranium (it allows for precise flux control in the core). CANDU PHWRs have been built in Canada, Argentina, China, India, Pakistan, Romania, and South Korea. India also operates a number of PHWRs, often termed 'CANDU-derivatives', built after the Government of Canada halted nuclear dealings with India following the 1974 Smiling Buddha nuclear weapon test.

Reaktor Bolshoy Moschnosti Kanalniy (High Power Channel Reactor) (RBMK)

Figure: *The Ignalina Nuclear Power Plant — a RBMK type (closed 2009)*

A Soviet design, built to produce plutonium as well as power. RBMKs are water cooled with a graphite moderator. RBMKs are in some respects similar to CANDU in that they are refuelable during power operation and employ a pressure tube design instead of a PWR-style pressure vessel.

However, unlike CANDU they are very unstable and large, making containment buildings for them expensive. A series of critical safety flaws have also been identified with the RBMK design, though some of these were corrected following the Chernobyl disaster. Their main attraction is their use of light water and un-enriched uranium. As of 2010, 11 remain open, mostly due to safety improvements and help from international safety agencies such as the DOE. Despite these safety improvements, RBMK reactors are still considered one of the most dangerous reactor designs in use. RBMK reactors were deployed only in the former Soviet Union.

Gas-Cooled Reactor (GCR) and Advanced Gas-Cooled Reactor (AGR)

Figure: *The Magnox Sizewell A nuclear power station*

Figure: *The Torness nuclear power station — an AGR*

These are generally graphite moderated and CO_2 cooled. They can have a high thermal efficiency compared with PWRs due to higher operating temperatures.

There are a number of operating reactors of this design, mostly in the United Kingdom, where the concept was developed. Older designs (i.e. Magnox stations) are either shut down or will be in the near future. However, the AGCRs have an anticipated life of a further 10 to 20 years. This is a thermal neutron reactor design. Decommissioning costs can be high due to large volume of reactor core.

Liquid-Metal Fast-Breeder Reactor (LMFBR)

This is a reactor design that is cooled by liquid metal, totally unmoderated, and produces more fuel than it consumes. They are said to "breed" fuel, because they produce fissionable fuel during operation because of neutron capture. These reactors can function much like a PWR in terms of efficiency, and do not require much high-pressure containment, as the liquid metal does not need to be kept at high pressure, even at very high temperatures. BN-350 and BN-600 in USSR and Superphénix in France were a reactor of this type, as was Fermi-I in the United States. The Monju reactor in Japan suffered a sodium leak in 1995 and was restarted in May 2010. All of them use/used liquid sodium. These reactors are fast neutron, not thermal neutron designs. These reactors come in two types:

Figure: *The Superphénix, one of the few FBRs*

Lead-Cooled

Using lead as the liquid metal provides excellent radiation shielding, and allows for operation at very high temperatures. Also, lead is (mostly) transparent to neutrons, so fewer neutrons are lost in the coolant, and the coolant does not become radioactive. Unlike sodium, lead is mostly inert, so there is less risk of explosion or accident, but such large quantities of lead may be problematic from toxicology and disposal points of view. Often a reactor of this type would use a lead-bismuth eutectic mixture. In this case, the bismuth would present some minor radiation problems, as it is not quite as transparent to neutrons, and can be transmuted to a radioactive isotope more readily than lead. The Russian Alfa class submarine uses a lead-bismuth-cooled fast reactor as its main power plant.

Sodium-Cooled

Most LMFBRs are of this type. The sodium is relatively easy to obtain and work with, and it also manages to actually prevent corrosion on the various reactor parts immersed in it. However, sodium explodes violently when exposed to water, so care must be taken, but such explosions would not be vastly more violent than (for example) a leak of superheated fluid from a SCWR or PWR. EBR-I, the first reactor to have a core meltdown, was of this type.

Pebble-Bed Reactors (PBR)

These use fuel molded into ceramic balls, and then circulate gas through the balls. The result is an efficient, low-maintenance, very safe reactor with inexpensive, standardized fuel. The prototype was the AVR.

Molten Salt Reactors

These dissolve the fuels in fluoride salts, or use fluoride salts for coolant. These have many safety features, high efficiency and a high power density suitable for vehicles. Notably, they have no high pressures or flammable components in the core. The prototype was the MSRE, which also used Thorium's fuel cycle to produce 0.1% of the radioactive waste of standard reactors.

Aqueous Homogeneous Reactor (AHR)

These reactors use soluble nuclear salts dissolved in water and mixed with a coolant and a neutron moderator.

Future and Developing Technologies

Advanced Reactors: More than a dozen advanced reactor designs are in various stages of development. Some are evolutionary from the PWR, BWR and PHWR designs above, some are more radical departures. The former include the advanced boiling water reactor (ABWR), two of which are now operating with others under construction, and the planned passively safe Economic Simplified Boiling Water Reactor (ESBWR) and AP1000 units.

- The Integral Fast Reactor (IFR) was built, tested and evaluated during the 1980s and then retired under the Clinton administration in the 1990s due to nuclear non-proliferation policies of the administration. Recycling spent fuel is the core of its design and it therefore produces only a fraction of the waste of current reactors.

- The pebble-bed reactor, a high-temperature gas-cooled reactor (HTGCR), is designed so high temperatures reduce power output by Doppler broadening of the fuel's neutron cross-section. It uses ceramic fuels so its safe operating temperatures exceed the power-reduction temperature range. Most designs are cooled by inert helium. Helium is not subject to steam explosions, resists neutron absorption leading to radioactivity, and does not dissolve contaminants that can become radioactive. Typical designs have more layers (up to 7) of passive containment than light water reactors (usually 3). A unique feature that may aid safety is that the fuel-balls actually form the core's mechanism, and are replaced one-by-one as they age. The design of the fuel makes fuel reprocessing expensive.
- The Small, sealed, transportable, autonomous reactor (SSTAR) is being primarily researched and developed in the US, intended as a fast breeder reactor that is passively safe and could be remotely shut down in case the suspicion arises that it is being tampered with.
- The Clean And Environmentally Safe Advanced Reactor (CAESAR) is a nuclear reactor concept that uses steam as a moderator – this design is still in development.
- The Reduced moderation water reactor builds upon the Advanced boiling water reactor(ABWR) that is presently in use, it is not a complete fast reactor instead using mostly epithermal neutrons, which are between thermal and fast neutrons in speed.
- The hydrogen-moderated self-regulating nuclear power module (HPM) is a reactor design emanating from the Los Alamos National Laboratory that uses uranium hydride as fuel.
- Subcritical reactors are designed to be safer and more stable, but pose a number of engineering and economic difficulties. One example is the Energy amplifier.
- Thorium-based reactors. It is possible to convert Thorium-232 into U-233 in reactors specially designed for the purpose. In this way, thorium, which is four times more abundant than uranium, can be used to breed U-233 nuclear fuel. U-233 is also believed to have favourable nuclear properties as compared to traditionally used U-235, including better neutron economy and lower production of long lived transuranic waste.
    - Advanced heavy-water reactor (AHWR)— A proposed heavy water moderated nuclear power reactor that will be the next generation design of the PHWR type. Under development in the Bhabha Atomic Research Centre (BARC), India.

- o KAMINI — A unique reactor using Uranium-233 isotope for fuel. Built in India by BARC and Indira Gandhi Centree for Atomic Research (IGCAR).
- o India is also planning to build fast breeder reactors using the thorium – Uranium-233 fuel cycle. The FBTR (Fast Breeder Test Reactor) in operation at Kalpakkam (India) uses Plutonium as a fuel and liquid sodium as a coolant.

Generation IV Reactors

Generation IV reactors are a set of theoretical nuclear reactor designs currently being researched. These designs are generally not expected to be available for commercial construction before 2030. Current reactors in operation around the world are generally considered second- or third-generation systems, with the first-generation systems having been retired some time ago. Research into these reactor types was officially started by the Generation IV International Forum (GIF) based on eight technology goals. The primary goals being to improve nuclear safety, improve proliferation resistance, minimize waste and natural resource utilisation, and to decrease the cost to build and run such plants.

- Gas-cooled fast reactor
- Lead-cooled fast reactor
- Molten salt reactor
- Sodium-cooled fast reactor
- Supercritical water reactor
- Very-high-temperature reactor

Generation V+ Reactors

Generation V reactors are designs which are theoretically possible, but which are not being actively considered or researched at present. Though such reactors could be built with current or near term technology, they trigger little interest for reasons of economics, practicality, or safety.

- Liquid-core reactor. A closed loop liquid-core nuclear reactor, where the fissile material is molten uranium or uranium solution cooled by a working gas pumped in through holes in the base of the containment vessel.
- Gas-core reactor. A closed loop version of the nuclear lightbulb rocket, where the fissile material is gaseous uranium-hexafluoride contained in a fused silica vessel. A working gas (such as hydrogen) would flow around this vessel and absorb the UV light produced by the reaction. In theory, using UF_6 as

a working fuel directly (rather than as a stage to one, as is done now) would mean lower processing costs, and very small reactors. In practice, running a reactor at such high power densities would probably produce unmanageable neutron flux, weakening most reactor materials, and therefore as the flux would be similar to that expected in fusion reactors, it would require similar materials to those selected by the International Fusion Materials Irradiation Facility.

  - o Gas core EM reactor. As in the gas core reactor, but with photovoltaic arrays converting the UV light directly to electricity.

- Fission fragment reactor
- Hybrid nuclear fusion. Would use the neutrons emitted by fusion to fission a blanket of fertile material, like U-238 or Th-232 and transmutate other reactor's spent nuclear fuel/nuclear waste into relatively more benign isotopes.

Fusion Reactors

Controlled nuclear fusion could in principle be used in fusion power plants to produce power without the complexities of handling actinides, but significant scientific and technical obstacles remain. Several fusion reactors have been built, but only recently reactors have been able to release more energy than the amount of energy used in the process. Despite research having started in the 1950s, no commercial fusion reactor is expected before 2050. The ITER project is currently leading the effort to commercialize fusion power.

Nuclear Fuel Cycle

Thermal reactors generally depend on refined and enriched uranium. Some nuclear reactors can operate with a mixture of plutonium and uranium. The process by which uranium ore is mined, processed, enriched, used, possibly reprocessed and disposed of is known as the nuclear fuel cycle.

Under 1% of the uranium found in nature is the easily fissionable U-235 isotope and as a result most reactor designs require enriched fuel. Enrichment involves increasing the percentage of U-235 and is usually done by means of gaseous diffusion or gas centrifuge. The enriched result is then converted into uranium dioxide powder, which is pressed and fired into pellet form. These pellets are stacked into tubes which are then sealed and called fuel rods. Many of these fuel rods are used in each nuclear reactor.

Most BWR and PWR commercial reactors use uranium enriched to about 4% U-235, and some commercial reactors with a high neutron economy do not require the fuel to be enriched at all (that is, they can use natural uranium). According to the International Atomic Energy Agency there are at least 100 research reactors in the world fuelled by highly enriched (weapons-grade/90% enrichment uranium). Theft risk of this fuel (potentially used in the production of a nuclear weapon) has led to campaigns advocating conversion of this type of reactor to low-enrichment uranium (which poses less threat of proliferation).

Fissile U-235 and non-fissile but fissionable and fertile U-238 are both used in the fission process. U-235 is fissionable by thermal (i.e. slow-moving) neutrons. A thermal neutron is one which is moving about the same speed as the atoms around it. Since all atoms vibrate proportionally to their absolute temperature, a thermal neutron has the best opportunity to fission U-235 when it is moving at this same vibrational speed. On the other hand, U-238 is more likely to capture a neutron when the neutron is moving very fast. This U-239 atom will soon decay into plutonium-239, which is another fuel. Pu-239 is a viable fuel and must be accounted for even when a highly enriched uranium fuel is used. Plutonium fissions will dominate the U-235 fissions in some reactors, especially after the initial loading of U-235 is spent. Plutonium is fissionable with both fast and thermal neutrons, which make it ideal for either nuclear reactors or nuclear bombs.

Most reactor designs in existence are thermal reactors and typically use water as a neutron moderator (moderator means that it slows down the neutron to a thermal speed) and as a coolant. But in a fast breeder reactor, some other kind of coolant is used which will not moderate or slow the neutrons down much. This enables fast neutrons to dominate, which can effectively be used to constantly replenish the fuel supply. By merely placing cheap unenriched uranium into such a core, the non-fissionable U-238 will be turned into Pu-239, "breeding" fuel.

In thorium fuel cycle thorium-232 absorbs a neutron in either a fast or thermal reactor. The thorium-233 beta decays to protactinium-233 and then to uranium-233, which in turn is used as fuel. Hence, like uranium-238, thorium-232 is a fertile material.

Fuelling of Nuclear Reactors

The amount of energy in the reservoir of nuclear fuel is frequently expressed in terms of "full-power days," which is the number of 24-hour periods (days) a reactor is scheduled for operation at full power output for the generation of heat energy. The number of full-power days in a reactor's operating cycle (between refuelling outage times) is related to the amount of fissile uranium-235 (U-235) contained in the fuel assemblies at the beginning of the cycle. A higher percentage of U-235 in the core at the beginning of a cycle will permit the reactor to be run for a greater number of full-power days.

At the end of the operating cycle, the fuel in some of the assemblies is "spent" and is discharged and replaced with new (fresh) fuel assemblies, although in practice it is the buildup of reaction poisons in nuclear fuel that determines the lifetime of nuclear fuel in a reactor. Long before all possible fission has taken place, the buildup of long-lived neutron absorbing fission byproducts impedes the chain reaction. The fraction of the reactor's fuel core replaced during refuelling is typically one-fourth for a boiling-water reactor and one-third for a pressurized-water reactor. The disposition and storage of this spent fuel is one of the most challenging aspects of the operation of a commercial nuclear power plant. This nuclear waste is highly radioactive and its toxicity presents a danger for thousands of years.

Not all reactors need to be shut down for refuelling; for example, pebble bed reactors, RBMK reactors, molten salt reactors, Magnox, AGR and CANDU reactors allow fuel to be shifted through the reactor while it is running. In a CANDU reactor, this also allows individual fuel elements to be situated within the reactor core that are best suited to the amount of U-235 in the fuel element.

The amount of energy extracted from nuclear fuel is called its burnup, which is expressed in terms of the heat energy produced per initial unit of fuel weight. Burn up is commonly expressed as megawatt days thermal per metric ton of initial heavy metal.

Fast-Neutron Reactor

A fast neutron reactor or simply a fast reactor is a category of nuclear reactor in which the fission chain reaction is sustained by fast neutrons. Such a reactor needs no neutron moderator, but must use fuel that is relatively rich in fissile material when compared to that required for a thermal reactor.

Neutrons released in fission events typically have energies greater than 1 MeV. The fission cross-sections for fissile materials, such as U-235, are greatest at much lower energies around 1 eV. The majority of nuclear reactors in operation are known as thermal reactors, which slow the high-energy ('fast') neutrons down to low ('thermal') energies, which are in thermal equilibrium with the reactor materials. This is achieved through elastic scattering of neutrons in a moderator. In a fast reactor, this process is avoided by eliminating any moderator and the fuel consists of materials with relatively large high-energy fission cross sections.

Advantages

Actinides and fission products by half-life

- *v*
- *t*
- *e*

Actinides by decay chain				***Half-life range (a)***	***Fission products of ^{235}U by yield***			
4*n*	4*n*+1	4*n*+2	4*n*+3			4.5–7%	0.04–1.25%	<0.001%
^{228}Ra№				4–6	†		^{155}Euþ	
^{244}Cm	^{241}Pu*ƒ*	^{250}Cf	^{227}Ac№	10–29		^{90}Sr	^{85}Kr	^{113m}Cdþ
^{232}U*ƒ*		^{238}Pu	^{243}Cm*ƒ*	29–97		^{137}Cs	^{151}Smþ	^{121m}Sn
^{248}Bk[3]	^{249}Cf*ƒ*	^{242m}Am*ƒ*		141–351	No fission products have a half-life in the range of 100–210k years...			
	^{241}Am		^{251}Cf*ƒ*[4]	430–900				
		^{226}Ra№	^{247}Bk	1.3k–1.6k				
^{240}Pu	^{229}Th	^{246}Cm	^{243}Am	4.7k–7.4k				
	^{245}Cm*ƒ*	^{250}Cm		8.3k–8.5k				
			^{239}Pu*ƒ*	24.1k				
		^{230}Th№	^{231}Pa№	32k–76k				
^{236}Np*ƒ*	^{233}U*ƒ*	^{234}U№		150k–250k	‡	^{99}Tc	^{126}Sn	
^{248}Cm		^{242}Pu		327k–375k			^{79}Se	
				1.53M		^{93}Zr		
	^{237}Np			2.1M–6.5M		^{135}Cs	^{107}Pd	
^{236}U			^{247}Cm*ƒ*	15M–24M			^{129}I	
^{244}Pu№				80M	...nor beyond 15.7M			
^{232}Th№		^{238}U№	^{235}U*ƒ*№	0.7G–14.1G				

Legend for superscript symbols
 has thermal neutron capture cross section in the range of 8–50 barns
ƒ fissile
m metastable isomer
№ naturally occurring radioactive material (NORM)
þ neutron poison (thermal neutron capture cross section greater than 3k barns)
† range 4a–97a: Medium-lived fission product
‡ over 200ka: Long-lived fission product

Legend for superscript symbols¡ has thermal neutron capture cross section in the range of 8–50 barns*f* fissilem metastable isomer[1] naturally occurring radioactive material (NORM)þ neutron poison (thermal neutron capture cross section greater than 3k barns)† range 4a–97a: Medium-lived fission product‡ over 200ka: Long-lived fission product.

Fast neutron reactors can reduce the total radiotoxicity of nuclear waste, and dramatically reduce the waste's lifetime. They can also use all or almost all of the fuel in the waste. Fast neutrons have an advantage in the transmutation of nuclear waste. With fast neutrons, the ratio between splitting and the capture of neutrons of plutonium or minor actinide is often larger than when the neutrons are slower, at thermal or near-thermal "epithermal" speeds. The transmuted odd-numbered actinides (e.g. from Pu-240 to Pu-241) split more easily. After they split, the actinides become a pair of "fission products." These elements have less total radiotoxicity. Since disposal of the fission products is dominated by the most radiotoxic fission product, Cesium 137, which has a half life of 30.1 years, the result is to reduce nuclear waste lifetimes from tens of millennia (from transuranic isotopes) to a few centuries. The processes are not perfect, but the remaining transuranics are reduced from a significant problem to a tiny percentage of the total waste, because most transuranics can be used as fuel.

- Fast reactors technically solve the "fuel shortage" argument against uranium-fuelled reactors without assuming unexplored reserves, or extraction from dilute sources such as ordinary granite or the ocean. They permit nuclear fuels to be bred from almost all the actinides, including known, abundant sources of depleted uranium and thorium, and light water reactor wastes. On average, more neutrons per fission are produced from fissions caused by fast neutrons than from those caused by thermal neutrons. This results in a larger surplus of neutrons beyond those required to sustain the chain reaction. These neutrons can be used to produce extra fuel, or to transmute long half-life waste to less troublesome isotopes, such as was done at the Phénix reactor in Marcoule in France, or some can be used for each purpose. Though conventional thermal reactors also produce excess neutrons, fast reactors can produce enough of them to breed more fuel than they consume. Such designs are known as fast breeder reactors.
- The fast reactor doesn't just transmute the inconvenient even-numbered transuranic elements (notably Pu-240 and U-238). It transmutes them, and then fissions them for power, so that these former wastes would actually become valuable.

Disadvantages

- Breeder reactors are costly to build and operate, and are not likely to be cost-competitive with thermal reactors unless the price of uranium increases dramatically.
- Due to the low cross sections of most materials at high neutron energies, critical mass in a fast reactor is much higher than a thermal reactor. In practice, this means significantly higher enrichment: >20% enrichment in a fast reactor compared to <5% enrichment in typical thermal reactors. This raises greater Nuclear proliferation and nuclear security issues.
- Sodium is often used as a coolant in fast reactors, because it does not moderate neutron speeds much and has a high heat capacity. However, it burns and foams in air. It has caused difficulties in reactors (e.g. USS Seawolf (SSN-575), Monju), although some sodium-cooled fast reactors have operated safely (notably the Superphénix and EBR-II for 30 years).
- Since liquid metals have low moderating power and ratio and no other moderator is present, the primary interaction of neutrons with liquid metal coolants is the (n,gamma) reaction, which induces radioactivity in the coolant. Boiling in the coolant, e.g. in an accident, would reduce coolant density and thus the absorption rate, such that the reactor has a positive void coefficient, which is dangerous and undesirable from a safety and accident standpoint. This can be avoided with a gas cooled reactor, since voids do not form in such a reactor during an accident; however, activation in the coolant remains a problem. A helium-cooled reactor would avoid this, since the elastic scattering and total cross sections are approximately equal, i.e. there are very few (n,gamma) reactions in the coolant and the low density of helium at typical operating conditions means that the amount neutrons have few interactions with coolant.

Nuclear Reactor Design

Coolant: Water, the most common coolant in thermal reactors, is generally not a feasible coolant for a fast reactor, because it acts as a neutron moderator. However the Generation IV reactor known as the supercritical water reactor with decreased coolant density may reach a hard enough neutron spectrum to be considered a fast reactor. Breeding, which is the primary advantage of fast over thermal reactors, may be accomplished with a thermal, light-water cooled & moderated system using very high enriched (~90%) uranium.

All current fast reactors are liquid metal cooled reactors. The early Clementine reactor used mercury coolant and plutonium metal fuel. NaK coolant is popular in test reactors due to its low melting point. In addition to its toxicity to humans, mercury has a high cross section for the (n,gamma) reaction, causing activation in the coolant and losing neutrons that could otherwise be absorbed in the fuel, which is why it is no longer used or considered as a coolant in reactors. Molten lead cooling has been used in naval propulsion units as well as some other prototype reactors. All large-scale fast reactors have used molten sodium coolant.

Another proposed fast reactor is a Molten Salt Reactor, one in which the molten salt's moderating properties are insignificant. This is typically achieved by replacing the light metal fluorides (e.g. LiF, BeF_2) in the salt carrier with heavier metal chlorides (e.g., KCl, RbCl, $ZrCl_4$).

Gas-cooled fast reactors have been the subject of research as well, as helium, the most commonly proposed coolant in such a reactor, has small absorption and scattering cross sections, thus preserving the fast neutron spectrum without significant neutron absorption in the coolant.

Nuclear Fuel

In practice, sustaining a fission chain reaction with fast neutrons means using relatively highly enriched uranium or plutonium. The reason for this is that fissile reactions are favoured at thermal energies, since the ratio between the Pu239 fission cross section and U238 absorption cross section is ~100 in a thermal spectrum and 8 in a fast spectrum. Fission and absorption cross sections are low for both Pu239 and U238 at high (fast) energies, which means that fast neutrons are likelier to pass through fuel without interacting than thermal neutrons; thus, more fissile material is needed. Therefore it is impossible to build a fast reactor using only natural uranium fuel. However, it is possible to build a fast reactor that will *breed* fuel (from fertile material) by producing more fissile material than it consumes. After the initial fuel charge such a reactor can be refuelled by reprocessing. Fission products can be replaced by adding natural or even depleted uranium with no further enrichment required. This is the concept of the fast breeder reactor or FBR.

So far, most fast neutron reactors have used either MOX (mixed oxide) or metal alloy fuel. Soviet fast neutron reactors have been using (high U-235 enriched) uranium fuel. The Indian prototype reactor has been using uranium-carbide fuel.

While criticality at fast energies may be achieved with uranium enriched to 5.5 weight percent Uranium-235, fast reactor designs have often been proposed with enrichments in the range of 20 percent for a variety of reasons, including core lifetime: If a fast reactor were loaded with the minimal critical mass, then the reactor would become subcritical after the first fission had occurred. Rather, an excess of fuel is inserted with reactivity control mechanisms, such that the reactivity control is inserted fully at the beginning of life to bring the reactor from supercritical to critical; as the fuel is depleted, the reactivity control is withdrawn to mitigate the negative reactivity feedback from fuel depletion and fission product poisons. In a fast breeder reactor, the above applies, though the reactivity from fuel depletion is also compensated by the breeding of either Uranium-233 or Plutonium-239 and 241 from Thorium 232 or Uranium 238, respectively.

Control

Like thermal reactors, fast neutron reactors are controlled by keeping the criticality of the reactor reliant on delayed neutrons, with gross control from neutron-absorbing control rods or blades.

They cannot, however, rely on changes to their moderators because there is no moderator. So Doppler broadening in the moderator, which affects thermal neutrons, does not work, nor does a negative void coefficient of the moderator. Both techniques are very common in ordinary light water reactors.

Doppler broadening from the molecular motion of the fuel, from its heat, can provide rapid negative feedback. The molecular movement of the fissionables themselves can tune the fuel's relative speed away from the optimal neutron speed. Thermal expansion of the fuel itself can also provide quick negative feedback. Small reactors such as those used in submarines may use doppler broadening or thermal expansion of neutron reflectors.

History

A 2008 IAEA proposal for a Fast Reactor Knowledge Preservation System notes that:

during the past 15 years there has been stagnation in the development of fast reactors in the industrialized countries that were involved, earlier, in intensive development of this area. All studies on fast reactors have been stopped in countries such as Germany, Italy, the United Kingdom and the United States of America and the only work being carried out is related to the decommissioning of fast reactors.

Many specialists who were involved in the studies and development work in this area in these countries have already retired or are close to retirement. In countries such as France, Japan and the Russian Federation that are still actively pursuing the evolution of fast reactor technology, the situation is aggravated by the lack of young scientists and engineers moving into this branch of nuclear power.

Figure: *Shevchenko BN350 desalination unit. View of the only nuclear-heated desalination unit in the world*

Nuclear Safety and Security

Nuclear safety and security covers the actions taken to prevent nuclear and radiation accidents or to limit their consequences. This covers nuclear power plants as well as all other nuclear facilities, the transportation of nuclear materials, and the use and storage of nuclear materials for medical, power, industry, and military uses.

The nuclear power industry has improved the safety and performance of reactors, and has proposed new and safer reactor designs. However, a perfect safety cannot be guaranteed. Potential sources of problems include human errors and external events that have a greater impact than anticipated: The designers of reactors at Fukushima in Japan did not anticipate that a tsunami generated by an earthquake would disable the backup systems that were supposed to stabilize the reactor after the earthquake. According to UBS AG, the Fukushima I nuclear accidents have cast doubt on whether even an advanced economy like Japan can master nuclear safety. Catastrophic scenarios involving terrorist attacks, insider sabotage, plowhshares actions and cyberattacks are also conceivable.

Figure: *2000 candles in memory of the Chernobyl disaster in 1986, at a commemoration 25 years after the nuclear accident, as well as for the Fukushima nuclear disaster of 2011.*

In his book, *Normal accidents*, Charles Perrow says that multiple and unexpected failures are built into society's complex and tightly-coupled nuclear reactor systems. Such accidents are unavoidable and cannot be designed around. To date, there have been three serious accidents (core damage) in the world since 1970, involving five reactors (one at Three Mile Island in 1979; one at Chernobyl in 1986; and three at Fukushima-Daiichi in 2011), corresponding to the beginning of the operation of generation II reactors.

Nuclear weapon safety, as well as the safety of military research involving nuclear materials, is generally handled by agencies different from those that oversee civilian safety, for various reasons, including secrecy. There are ongoing concerns about terrorist groups acquiring nuclear bomb-making material.

Overview of Nuclear Processes and Safety Issues

As of 2011, nuclear safety considerations occur in a number of situations, including:

- Nuclear fission power used in nuclear power stations, and nuclear submarines and ships
- Nuclear weapons
- Fissionable fuels such as uranium and plutonium and their extraction, storage and use

- Radioactive materials used for medical, diagnostic, batteries for some space projects, and research purposes
- Nuclear waste, the radioactive waste residue of nuclear materials
- Nuclear fusion power, a technology under long-term development
- Unplanned entry of nuclear materials into the biosphere and food chain (living plants, animals and humans) if breathed or ingested.

With the exception of thermonuclear weapons and experimental fusion research, all safety issues specific to nuclear power stems from the need to limit the biological uptake of committed dose (toxicity), and effective dose due to the radioactivity of heavy fissionable materials, waste byproducts, and from the risks of unplanned or uncontrolled nuclear accidents.

Nuclear safety therefore covers at minimum: -

- Extraction, transportation, storage, processing, and disposal of fissionable materials
- Safety of nuclear power generators
- Control and safe management of nuclear weapons, nuclear material capable of use as a weapon, and other radioactive materials
- Safe handling, accountability and use in industrial, medical and research contexts
- Disposal of nuclear waste
- Limitations on exposure to radiation

Responsible Agencies

Internationally the International Atomic Energy Agency "works with its Member States and multiple partners worldwide to promote safe, secure and peaceful nuclear technologies." Some scientists say that the 2011 Japanese nuclear accidents have revealed that the nuclear industry lacks sufficient oversight, leading to renewed calls to redefine the mandate of the IAEA so that it can better police nuclear power plants worldwide. There are several problems with the IAEA says Najmedin Meshkati of University of Southern California:

It recommends safety standards, but member states are not required to comply; it promotes nuclear energy, but it also monitors nuclear use; it is the sole global organisation overseeing the nuclear energy industry, yet it is also weighed down by checking compliance with the Nuclear Non-Proliferation Treaty (NPT).

Figure: *IAEA headquarters in Vienna, Austria*

Many nations utilising nuclear power have special institutions overseeing and regulating nuclear safety. Civilian nuclear safety in the U.S. is regulated by the Nuclear Regulatory Commission (NRC). However, critics of the nuclear industry complain that the regulatory bodies are too intertwined with the inustries themselves to be effective. The book *The Doomsday Machine* for example, offers a series of examples of national regulators, as they put it 'not regulating, just waving' (a pun on *waiving*) to argue that, in Japan, for example, "regulators and the regulated have long been friends, working together to offset the doubts of a public brought up on the horror of the nuclear bombs". Other examples offered include:

- in the United States, a dangerous custom whereby only supporters of the nuclear industry are allowed to supervise it and lobbyists have been allowed to have an effective veto over regulators.
- in China, where Kang Rixin, former general manager of the state-owned China National Nuclear Corporation, was sentenced to life in jail in 2010 for accepting bribes (and other abuses), a verdict raising questions about the quality of his work on the safety and trustworthiness of China's nuclear reactors.

- in India, where the nuclear regulator reports to the national Atomic Energy Commission, which champions the building of nuclear power plants there and the chairman of the Atomic Energy Regulatory Board, S. S. Bajaj, was previously a senior executive at the Nuclear Power Corporation of India, the company he is now helping to regulate.
- in Japan, where the regulator reports to the Ministry of Economy, Trade and Industry, which overtly seeks to promote the nuclear industry and ministry posts and top jobs in the nuclear business are passed among the same small circle of experts.

The book argues that nuclear safety is compromised by the suspicion that, as Eisaku Sato, formerly a governor of Fukushima province (with its infamous nuclear reactor complex), has put it of the regulators: "They're all birds of a feather".

The safety of nuclear plants and materials controlled by the U.S. government for research, weapons production, and those powering naval vessels is not governed by the NRC. In the UK nuclear safety is regulated by the Office for Nuclear Regulation (ONR) and the Defence Nuclear Safety Regulator (DNSR). The Australian Radiation Protection and Nuclear Safety Agency (ARPANSA) is the Federal Government body that monitors and identifies solar radiation and nuclear radiation risks in Australia. It is the main body dealing with ionizing and non-ionizing radiation and publishes material regarding radiation protection.

Other agencies include:

- Autorité de sûreté nucléaire
- Canadian Nuclear Safety Commission
- Radiological Protection Institute of Ireland
- Federal Atomic Energy Agency in Russia
- Kernfysische dienst, (NL)
- Pakistan Nuclear Regulatory Authority
- Bundesamt für Strahlenschutz, (DE)
- Atomic Energy Regulatory Board (India)

Nuclear Power Plant

Complexity: Nuclear power plants are some of the most sophisticated and complex energy systems ever designed. Any complex system, no matter how well it is designed and engineered, cannot be deemed failure-proof. Veteran journalist and author Stephanie Cooke has argued:

The reactors themselves were enormously complex machines with an incalculable number of things that could go wrong. When that happened at Three Mile Island in 1979, another fault line in the nuclear world was exposed. One malfunction led to another, and then to a series of others, until the core of the reactor itself began to melt, and even the world's most highly trained nuclear engineers did not know how to respond. The accident revealed serious deficiencies in a system that was meant to protect public health and safety.

The 1979 Three Mile Island accident inspired Perrow's book *Normal Accidents*, where a nuclear accident occurs, resulting from an unanticipated interaction of multiple failures in a complex system. TMI was an example of a normal accident because it was "unexpected, incomprehensible, uncontrollable and unavoidable".

Perrow concluded that the failure at Three Mile Island was a consequence of the system's immense complexity. Such modern high-risk systems, he realised, were prone to failures however well they were managed. It was inevitable that they would eventually suffer what he termed a 'normal accident'. Therefore, he suggested, we might do better to contemplate a radical redesign, or if that was not possible, to abandon such technology entirely. .

A fundamental issue contributing to a nuclear power system's complexity is its extremely long lifetime. The timeframe from the start of construction of a commercial nuclear power station through the safe disposal of its last radioactive waste, may be 100 to 150 years.

Failure Modes of Nuclear Power Plants

There are concerns that a combination of human and mechanical error at a nuclear facility could result in significant harm to people and the environment:

Operating nuclear reactors contain large amounts of radioactive fission products which, if dispersed, can pose a direct radiation hazard, contaminate soil and vegetation, and be ingested by humans and animals. Human exposure at high enough levels can cause both short-term illness and death and longer-term death by cancer and other diseases.

It is impossible for a commercial nuclear reactor to explode like a nuclear bomb since the fuel is never sufficiently enriched for this to occur.

Nuclear reactors can fail in a variety of ways. Should the instability of the nuclear material generate unexpected behaviour, it may result in an uncontrolled power excursion. Normally, the cooling system in a reactor is designed to be able to handle the excess heat this causes; however, should the reactor also experience a loss-of-coolant accident, then the fuel may melt or cause the vessel in which it is contained to overheat and melt. This event is called a nuclear meltdown.

After shutting down, for some time the reactor still needs external energy to power its cooling systems. Normally this energy is provided by the power grid to which that plant is connected, or by emergency diesel generators. Failure to provide power for the cooling systems, as happened in Fukushima I, can cause serious accidents.

Nuclear safety rules in the United States "do not adequately weigh the risk of a single event that would knock out electricity from the grid and from emergency generators, as a quake and tsunami recently did in Japan", Nuclear Regulatory Commission officials said in June 2011.

Vulnerability of Nuclear Plants to Attack

Nuclear reactors become preferred targets during military conflict and, over the past three decades, have been repeatedly attacked during military air strikes, occupations, invasions and campaigns:

- In September 1980, Iran bombed the Al Tuwaitha nuclear complex in Iraq in Operation Scorch Sword.
- In June 1981, an Israeli air strike completely destroyed Iraq's Osirak nuclear research facility in Operation Opera.
- Between 1984 and 1987, Iraq bombed Iran's Bushehr nuclear plant six times.
- On 8 January 1982, Umkhonto we Sizwe, the armed wing of the ANC, attacked South Africa's Koeberg nuclear power plant while it was still under construction.
- In 1991, the U.S. bombed three nuclear reactors and an enrichment pilot facility in Iraq.
- In 1991, Iraq launched Scud missiles at Israel's Dimona nuclear power
- In September 2007, Israel bombed a Syrian reactor under construction.

In the U.S., plants are surrounded by a double row of tall fences which are electronically monitored. The plant grounds are patrolled by a sizeable force of armed guards. The NRC's "Design Basis Threat" criteria for plants is a secret, and so what size of attacking force the plants are able to protect against is unknown. However, to scram (make an emergency shutdown) a plant takes fewer than 5 seconds while unimpeded restart takes hours, severely hampering a terrorist force in a goal to release radioactivity.

Attack from the air is an issue that has been highlighted since the September 11 attacks in the U.S. However, it was in 1972 when three hijackers took control of a domestic passenger flight along the east coast of the U.S. and threatened to crash the plane into a U.S. nuclear weapons plant in Oak Ridge, Tennessee. The plane got as close as 8,000 feet above the site before the hijackers' demands were met.

The most important barrier against the release of radioactivity in the event of an aircraft strike on a nuclear power plant is the containment building and its missile shield. Current NRC Chairman Dale Klein has said "Nuclear power plants are inherently robust structures that our studies show provide adequate protection in a hypothetical attack by an airplane. The NRC has also taken actions that require nuclear power plant operators to be able to manage large fires or explosions—no matter what has caused them."

In addition, supporters point to large studies carried out by the U.S. Electric Power Research Institute that tested the robustness of both reactor and waste fuel storage and found that they should be able to sustain a terrorist attack comparable to the September 11 terrorist attacks in the U.S. Spent fuel is usually housed inside the plant's "protected zone" or a spent nuclear fuel shipping cask; stealing it for use in a "dirty bomb" would be extremely difficult. Exposure to the intense radiation would almost certainly quickly incapacitate or kill anyone who attempts to do so.

Plant Location

In many countries, plants are often located on the coast, in order to provide a ready source of cooling water for the essential service water system. As a consequence the design needs to take the risk of flooding and tsunamis into account. The World Energy Council (WEC) argues disaster risks are changing and increasing the likelihood of disasters such as earthquakes, cyclones, hurricanes, typhoons, flooding. High temperatures, low precipitation levels and severe droughts may lead to fresh water shortages. Seawater is corrosive and so nuclear energy supply is likely to be negatively affected by the fresh water shortage. This generic problem may become increasingly significant over time. Failure to calculate the risk of flooding correctly lead to a Level 2 event on the International Nuclear Event Scale during the 1999 Blayais Nuclear Power Plant flood, while flooding caused by the 2011 Tôhoku earthquake and tsunami lead to the Fukushima I nuclear accidents.

Figure: *Angra Nuclear Power Plant in Rio de Janeiro, Brazil.*

The design of plants located in seismically active zones also requires the risk of earthquakes and tsunamis to be taken into account. Japan, India, China and the USA are among the countries to have plants in earthquake-prone regions. Damage caused to Japan's Kashiwazaki-Kariwa Nuclear Power Plant during the 2007 Chfietsu offshore earthquake underlined concerns expressed by experts in Japan prior to the Fukushima accidents, who have warned of a *genpatsu-shinsai* (domino-effect nuclear power plant earthquake disaster).

Multiple Reactors

The Fukushima nuclear disaster illustrated the dangers of building multiple nuclear reactor units close to one another. This proximity triggered the parallel, chain-reaction accidents that led to hydrogen explosions damaging reactor buildings and water draining from open-air spent fuel pools — a situation that was potentially more dangerous than the loss of reactor cooling itself. Because of the closeness of the reactors, Plant Director Masao Yoshida "was put in the position of trying to cope simultaneously with core meltdowns at three reactors and exposed fuel pools at three units".

Nuclear Safety Systems

The three primary objectives of nuclear safety systems as defined by the Nuclear Regulatory Commission are to shut down the reactor, maintain it in a shutdown condition, and prevent the release of radioactive material during events and accidents. These objectives are accomplished using a variety of equipment, which is part of different systems, of which each performs specific functions.

Routine Emissions of Radioactive Materials

During everyday routine operations, emissions of radioactive materials from nuclear plants are released to the outside of the plants although they are quite slight amounts. The daily emissions go into the air, water and soil.

NRC says, "nuclear power plants sometimes release radioactive gases and liquids into the environment under controlled, monitored conditions to ensure that they pose no danger to the public or the environment", and "routine emissions during normal operation of a nuclear power plant are never lethal".

According to the United Nations (UNSCEAR), regular nuclear power plant operation including the nuclear fuel cycle amounts to 0.0002 mSv (milli-Sievert) annually in average public radiation exposure; the legacy of the Chernobyl disaster is 0.002 mSv/yr as a global average as of a 2008 report; and natural radiation exposure averages 2.4 mSv annually although frequently varying depending on an individual's location from 1 to 13 mSv.

Japanese Public Perception of Nuclear Power Safety

In March 2012, Prime Minister Yoshihiko Noda said that the Japanese government shared the blame for the Fukushima disaster, saying that officials had been blinded by an image of the country's technological infallibility and were "all too steeped in a safety myth."

Japan has been accused by authors such as journalist Yoichi Funabashi of having an "aversion to facing the potential threat of nuclear emergencies." According to him, a national program to develop robots for use in nuclear emergencies was terminated in midstream because it "smacked too much of underlying danger." Though Japan is a major power in robotics, it had none to send in to Fukushima during the disaster. He mentions that Japan's Nuclear Safety Commission stipulated in its safety guidelines for light-water nuclear facilities that "the potential for extended loss of power need not be considered."

However, this kind of extended loss of power to the cooling pumps caused the Fukushima meltdown.

Economics of Nuclear Power

Internationally the price of nuclear plants rose 15% annually in 1970-1990. Total costs rose tenfold. The nuclear plant construction time became douple. According to Al Gore if intended plan does not hold, the delay cost a billion dollars a year.

The economics of new nuclear power plants is a controversial subject, since there are diverging views on this topic, and multi-billion dollar investments ride on the choice of an energy source. Nuclear power plants typically have high capital costs for building the plant, but low fuel costs. Therefore, comparison with other power generation methods is strongly dependent on assumptions about construction timescales and capital financing for nuclear plants as well as the future costs of fossil fuels and renewables as well as for energy storage solutions for intermittent power sources. Cost estimates also need to take into account plant decommissioning and nuclear waste storage costs. On the other hand measures to mitigate global warming, such as a carbon tax or carbon emissions trading, may favour the economics of nuclear power.

In recent years there has been a slowdown of electricity demand growth and financing has become more difficult, which has an impact on large projects such as nuclear reactors, with very large upfront costs and long project cycles which carry a large variety of risks. In Eastern Europe, a number of long-established projects are struggling to find finance, notably Belene in Bulgaria and the additional reactors at Cernavoda in Romania, and some potential backers have pulled out. Where cheap gas is available and its future supply relatively secure, this also poses a major problem for nuclear projects.

Analysis of the economics of nuclear power must take into account who bears the risks of future uncertainties. To date all operating nuclear power plants were developed by state-owned or regulated utility monopolies where many of the risks associated with construction costs, operating performance, fuel price, accident liability and other factors were borne by consumers rather than suppliers. In addition, because the potential liability from a nuclear accident is so great, the full cost of liability insurance is generally limited/capped by the government, which the U.S. Nuclear Regulatory Commission concluded constituted a significant subsidy. Many countries have now liberalized the electricity market where these risks, and the risk of cheaper competitors emerging before capital costs are recovered, are borne by plant suppliers and operators rather than consumers, which leads to a significantly different evaluation of the economics of new nuclear power plants.

Following the 2011 Fukushima Daiichi nuclear disaster, costs are expected to increase for currently operating and new nuclear power plants, due to increased requirements for on-site spent fuel management and elevated design basis threats.

Magnetic Confinement Fusion

Magnetic confinement fusion is an approach to generating fusion power that uses magnetic fields (which is a magnetic influence of electric currents and magnetic materials) to confine the hot fusion fuel in the form of a plasma. Magnetic confinement is one of two major branches of fusion energy research, the other being inertial confinement fusion. The magnetic approach is more highly developed and is usually considered more promising for energy production. Construction of a 500-MW heat generating fusion plant using tokamak magnetic confinement geometry, the ITER, began in France in 2007.

Fusion reactions combine light atomic nuclei such as hydrogen to form heavier ones such as helium. In order to overcome the electrostatic repulsion between them, the nuclei must have a temperature of several tens of millions of degrees, under which conditions they no longer form neutral atoms but exist in the plasma state. In addition, sufficient density and energy confinement are required, as specified by the Lawson criterion.

Magnetic confinement fusion attempts to create the conditions needed for fusion energy production by using the electrical conductivity of the plasma to contain it with magnetic fields. The basic concept can be thought of in a fluid picture as a balance between magnetic pressure and plasma pressure, or in terms of individual particles spiraling along magnetic field lines.

The pressure achievable is usually on the order of one bar with a confinement time up to a few seconds. In contrast, inertial confinement has a much higher pressure but a much lower confinement time. Most magnetic confinement schemes also have the advantage of being more or less steady state, as opposed to the inherently pulsed operation of inertial confinement.

The simplest magnetic configuration is a solenoid, a long cylinder wound with magnetic coils producing a field with the lines of force running parallel to the axis of the cylinder. Such a field would hinder ions and electrons from being lost radially, but not from being lost from the ends of the solenoid.

There are two approaches to solving this problem. One is to try to stop up the ends with a magnetic mirror, the other is to eliminate the ends altogether by bending the field lines around to close on themselves. A simple toroidal field, however, provides poor confinement because the radial gradient of the field strength results in a drift in the direction of the axis.

Magnetic Mirrors

A major area of research in the early years of fusion energy research was the magnetic mirror. Most early mirror devices attempted to confine plasma near the focus of a non-planar magnetic field, or to be more precise, two such mirrors located close to each other and oriented at right angles. In order to escape the confinement area, nuclei had to enter a small annular area near each magnet. It was known that nuclei would escape through this area, but by adding and heating fuel continually it was felt this could be overcome. As development of mirror systems progressed, additional sets of magnets were added to either side, meaning that the nuclei had to escape through two such areas before leaving the reaction area entirely. A highly developed form, the Mirror Fusion Test Facility (MFTF), used two mirrors at either end of a solenoid to increase the internal volume of the reaction area.

Toroidal Machines

An early attempt to build a magnetic confinement system was the stellarator, introduced by Lyman Spitzer in 1951. Essentially the stellarator consists of a torus that has been cut in half and then attached back together with straight "crossover" sections to form a figure depicted earlier. This has the effect of propagating the nuclei from the inside to outside as it orbits the device, thereby canceling out the drift across the axis, at least if the nuclei orbit fast enough. Newer versions of the stellarator design have replaced the "mechanical" drift cancellation with additional magnets that "wind" the field lines into a helix to cause the same effect.

In 1968 Russian research on the toroidal tokamak was first presented in public, with results that far outstripped existing efforts from any competing design, magnetic or not. Since then the majority of effort in magnetic confinement has been based on the tokamak principle. In the tokamak a current is periodically driven through the plasma itself, creating a field "around" the torus that combines with the toroidal field to produce a winding field in some ways similar to that in a modern stellarator, at least in that nuclei move from the inside to the outside of the device as they flow around it.

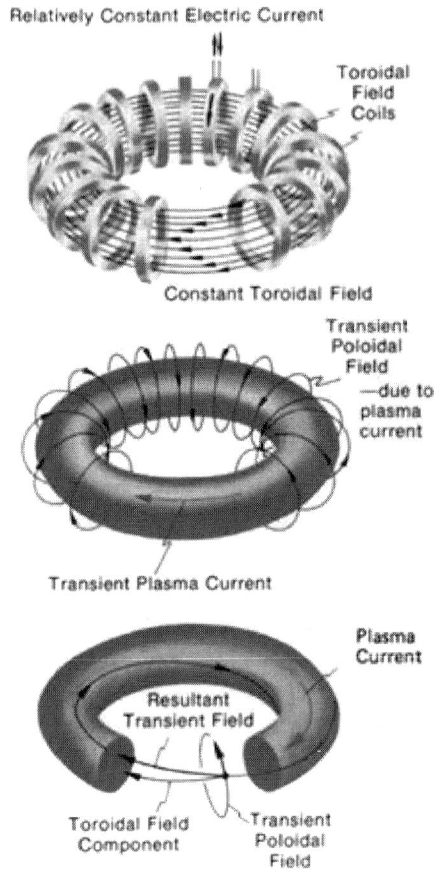

Figure: *Tokamak magnetic fields.*

In 1991, START was built at Culham, UK, as the first purpose built spherical tokamak. This was essentially a spheromak with an inserted central rod. START produced impressive results, with β values at approximately 40% - three times that produced by standard tokamaks at the time. The concept has been scaled up to higher plasma currents and larger sizes, with the experiments NSTX (US), MAST (UK) and Globus-M (Russia) currently running. Spherical tokamaks have improved stability properties compared to conventional tokamaks and as such the area is receiving considerable experimental attention. However spherical tokamaks to date have been at low toroidal field and as such are impractical for fusion neutron devices.

Some more novel configurations produced in toroidal machines are the reversed field pinch and the Levitated Dipole Experiment.

Compact Toroids

Compact toroids, e.g. the spheromak and the Field-Reversed Configuration, attempt to combine the good confinement of closed magnetic surfaces configurations with the simplicity of machines without a central core. An early experiment of this type was Trisops.

Magnetic Fusion Energy

All of these devices have faced considerable problems being scaled up and in their approach toward the Lawson criterion. One researcher has described the magnetic confinement problem in simple terms, likening it to squeezing a balloon – the air will always attempt to "pop out" somewhere else. Turbulence in the plasma has proven to be a major problem, causing the plasma to escape the confinement area, and potentially touch the walls of the container. If this happens, a process known as "*sputtering*", high-mass particles from the container (often steel and other metals) are mixed into the fusion fuel, lowering its temperature.

Progress has been remarkable – both in the significant progress toward a "burning" plasma and in the advance of scientific understanding. In 1997, scientists at the Joint European Torus (JET) facilities in the UK produced 16 megawatts of fusion power in the laboratory and have studied the behaviour of fusion products (alpha particles) in weakly burning plasmas. Underlying this progress are strides in fundamental understanding, which have led to the ability to control aspects of plasma behaviour.

For example, scientists can now exercise a measure of control over plasma turbulence and resultant energy leakage, long considered an unavoidable and intractable feature of plasmas; the plasma pressure above which the plasma disassembles can now be made large enough to sustain a fusion reaction rate acceptable for a power plant. Electromagnetic waves can be injected and steered to manipulate the paths of plasma particles and then to produce the large electrical currents necessary to produce the magnetic fields to confine the plasma. These and other control capabilities have flowed from advances in basic understanding of plasma science in such areas as plasma turbulence, plasma macroscopic stability, and plasma wave propagation. Much of this progress has been achieved with a particular emphasis on the tokamak.

Fusion Power

Fusion power is the energy generated by nuclear fusion processes. In fusion reactions, two light atomic nuclei fuse to form a heavier nucleus (in contrast with fission power). In doing so they release a comparatively large amount of energy arising from the binding energy due to the strong nuclear force that is manifested as an increase in temperature of the reactants. Fusion power is a primary area of research in plasma physics.

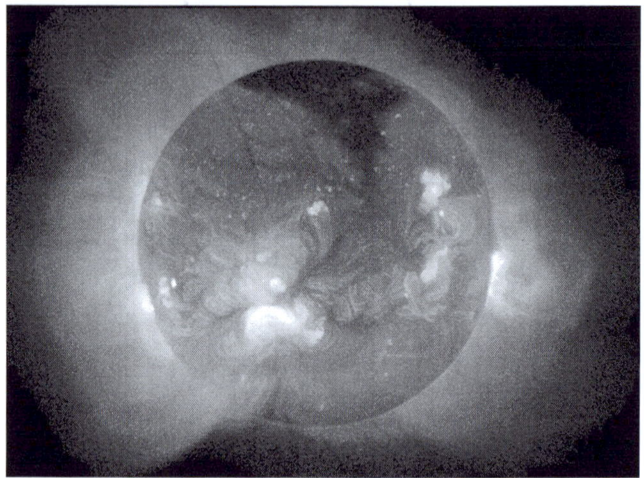

Figure: *The Sun is a natural fusion reactor.*

Fusion energy is released by stars - and generated all heavy atoms in the known universe. It is also released when hydrogen bombs - which use fission only to detonate a fusion explosion - explode. Despite the lack of commercial viability at present, many candidate power reactions are proven in labs. The fusion reactions all occur naturally inside stars and artificially inside bombs. Thus, there is no credible scientific objection to fusion power, it is clearly and only a technological, economic and social (use or abuse of the power released) set of issues. A reasonable concern is whether small fusion reactors could be used as explosive devices. This may be a factor steering institutional research to larger scale design.

Fusion power research seeks the commercial production of net usable power from a fusion source, similar to the usage of the term "steam power". The primary problem is confining the fusion reaction so it continues using its own energy. Leading designs for controlled fusion research use magnetic (tokamak design) or inertial (laser) confinement of a plasma. Both approaches are still under development and are years away from commercial operation. A few other models such as purely inertial confinement using centrifugal force are being explored for heavier fusion fuel such as liquid lithium, but these are not thought suitable for large scale power plants.

In most large scale commercial proposals, heat from the fusion reaction is used to operate a steam turbine that drives electrical generators, as in existing fossil fuel and nuclear fission power stations. Smaller scale reactors however often propose to use much simpler thermocouples - with no moving parts or hot liquids - to generate electricity from heat at a lower electrical efficiency, but with more safety and smaller size. Smaller fission reactors such as those on nuclear submarines have proven this approach, and research into materials suitable for fusion date back to at least 1981. Because of the variety of fusion reactions, the potential use of solid boron or liquid lithium fuel, a non-radioactive supply and minimally radioactive waste chain, smaller reactors may be deployed outside the military without these concerns, which have inhibited small uranium reactors.

However, fusion reactions do have their own radiation concerns. In small fusors, fast neutron radiation is released by the fusion reaction, but bremsstrahlung x-ray radiation caused by the slowing down of electrons is the primary human hazard.

If neutrons convey the energy released, then whatever medium absorbs the neutron scattering eventually becomes mildly radioactive and must also be disposed of, e.g. ordinary water becomes heavy water. In a fusion supply chain heavy water is actually useful again as a fuel, making a breeder reactor style of design more practical. However the overall complexity of this design has proven problematic in fission reactors such as the CANDU and may be uneconomic. It may be possible to retrofit turbines of existing fission reactors for thorium fission and for fusion-based neutron scatter.

However, aneutronic fusion and minimal neutron energy reactions such as proton-boron have received more research attention in the 2010s, largely for their potential to create small clean portable energy sources without neutron pollution.

Mechanism

Fusion happens when two (or more) nuclei come close enough for the strong nuclear force to exceed the electrostatic force and pull them together. This process takes light nuclei and forms a heavier one, through a nuclear reaction. For nuclei lighter than iron-56 this is exothermic and releases energy. For nuclei heavier than iron-56 this is endothermic and requires an external source of energy. Hence, nuclei smaller than iron-56 are more likely to fuse while those heavier than iron-56 are more likely to break apart.

To fuse, nuclei must overcome the repulsive Coulomb force. This is a force caused by the nuclei containing positively charged protons that repel via the electromagnetic force. To overcome this "Coulomb barrier", the atoms must have a high kinetic energy. There are several ways of doing this, including heating or acceleration. Once an atom is heated above its ionization energy, its electrons are stripped away, leaving just the bare nucleus: the ion. Most fusion experiments use a hot cloud of ions and electrons. This cloud is known as a plasma. Most fusion reactions produce neutrons, which can be detected and degrade materials.

Theoretically, any atom could be fused, if enough pressure and temperature was applied. Mankind has studied many high energy fusion reactions, using particles beams. These are fired at a target. For a power plant, however, we are currently limited to only the light elements. Hydrogen is ideal: because of its small charge, it is the easiest atom to fuse. This reaction produces helium.

Cross Section

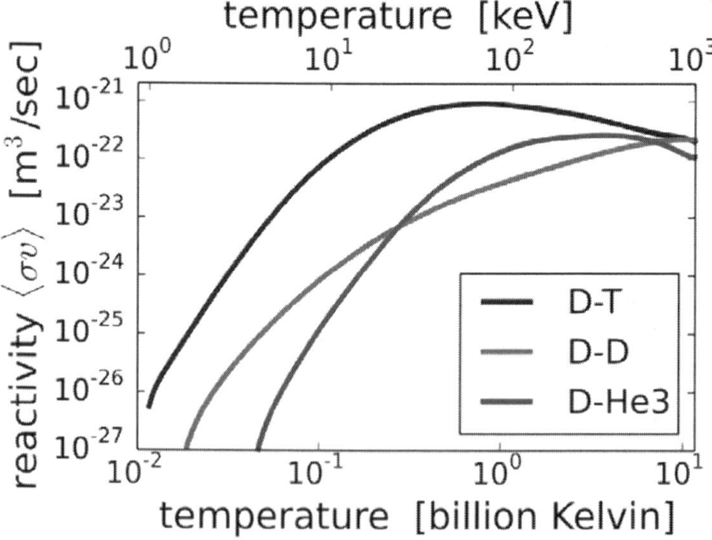

Figure: *The fusion reaction rate increases rapidly with temperature until it maximizes and then gradually drops off. The deuterium-tritium fusion rate peaks at a lower temperature (about 70 keV, or 800 million kelvin) and at a higher value than other reactions commonly considered for fusion energy.*

A reaction's cross section, (denoted ó) is the measure of how likely a fusion reaction will happen. It is a probability, and it depends on the velocity of the two nuclei when they strike one another. If the atoms move faster, fusion is more likely. If the atoms hit head on, fusion is more likely. Cross sections for many different fusion reactions were measured mainly in the 1970s using particle beams. A beam of species A was fired at species B at different speeds, and the amount of neutrons coming off was measured. Neutrons are a key product of fusion reactions. These nuclei are flying around in a hot cloud, with some distribution of velocities. If the plasma is thermalized, then the distribution looks like a bell curve, or maxwellian distribution. In this case, it is useful to take the average cross section over the velocity distribution. This is entered into the volumetric fusion rate:

$$P_{\text{fusion}} = n_A n_B \langle \sigma v_{A,B} \rangle E_{\text{fusion}}$$

where:

- P_{fusion} is the energy made by fusion, per time and volume
- n is the number density of species A or B, the particles in the volume
- $\langle \sigma v_{A,B} \rangle$ is the cross section of that reaction, average over all the velocities of the two species v
- E_{fusion} is the energy released by that fusion reaction.

Lawson Criterion

This equation shows that energy varies with the temperature, density, speed of collision, and fuel used. To reach net power, fusion reactions have to occur fast enough to make up for energy losses. Any power plant using fusion will hold in this hot cloud. Plasma clouds lose energy through conduction and radiation. Conduction is when ions, electrons or neutrals touch a surface and leak out. Energy is lost with the particle. Radiation is when energy leaves the cloud as light. Radiation increases as the temperature rises. To get net power from fusion, you must overcome these losses. This leads to an equation for power output.

$$P_{\text{out}} = \eta_{\text{capture}} \left(P_{\text{fusion}} - P_{\text{conduction}} - P_{\text{radiation}} \right)$$

where:

- η, is the efficiency with which the plant captures energy

John Lawson used this equation to estimate some conditions for net power based on a Maxwellian cloud. This is the Lawson criterion.

Density, Temperature, Time: ntô

The Lawson criterion argues that a machine holding in a hot, thermalized and quasi-neutral plasma, has to meet basic criteria to overcome the radiation losses, conduction losses and a power plant efficiency of 30 percent. This became known as the "triple product": the plasma density and temperature and how long it is held in. For many years, work has been focused on reaching the highest triple product possible. This emphasis on $(nt\tau)$ as a metric of success, has hurt other considerations like cost, size, complexity and efficiency. This has led to larger, more complicated and more expensive machines like ITER and NIF.

Energy Capture

There are several proposals for energy capture. The simplest is using a heat cycle to heat a fluid with fusion reactions. It has been proposed to use the neutrons generated by fusion to re-generate a spent fission fuel. In addition, direct energy conversion, has been developed (at LLNL in the 1980s) as a method to maintain a voltage using the products of a fusion reaction. This has demonstrated an energy capture efficiency of 48 percent.

Possible Approaches

Resonant cavity solves problem.. One approach, not yet explored, is to combine a tokamak with a tuned magnetic field such that the resonant cavity contains the temperature for a finite but very short time. During this period there will be a net energy gain.

Magnetic Confinement Fusion

Tokamak The most well developed and well funded approach to fusion energy. As of January 2011 there were an estimated 177 tokamak experiments either planned, decommissioned or currently operating, worldwide. This method races hot plasma around in a magnetically confined ring. When completed, ITER will be the world's largest Tokamak.

Stellarator These are twisted rings of hot plasma. Stellarators are distinct from tokamak in that they are not azimuthally symmetric. Instead, they have a discrete rotational symmetry, often fivefold, like a regular pentagon. Stellarators were developed by Lyman Spitzer in 1950. There are four designs: Torsatron, Heliotron, Heliac and Helias.

Levitated Dipole Experiment (LDX) These use a solid superconducting torus. This is magnetically levitated inside the reactor chamber. The superconductor forms an axisymmetric magnetic field that contains the plasma. The LDX was developed between MIT and Columbia University after 2000 by Jay Kesner and Michael E. Mauel.

Magnetic mirror Developed by Richard F. Post and teams at LLNL in the 1960s. Magnetic mirrors reflected hot plasma back and forth in a line. Variations included the magnetic bottle and the biconic cusp. A series of well-funded, large, mirror machines were built by the US government in the 1970s and 1980s.

Field-reversed configuration This device confines a plasma on closed magnetic field lines without a central penetration.

Inertial Confinement Fusion

Direct drive In this technique, lasers directly blast a pellet of fuel. The goal is to start ignition, a fusion chain reaction. Ignition was first suggested by John Nuckolls, in 1972. Notable direct drive experiments have been conducted at the Laboratory for Laser Energetics, Laser Mégajoule and the GEKKO XII facilities.

Fast ignition This method uses two laser blasts. The first blast compresses the fusion fuel, while the second high energy pulse ignites it. Experiments have been conducted at the Laboratory for Laser Energetics using the Omega and Omega EP systems.

Indirect drive In this technique, lasers blasts a structure around the pellet of fuel. This structure is known as a Hohlraum. As it disintegrates the pellet is bathed in a more uniform x-ray light, creating better compression. The largest system using this method is the National Ignition Facility.

Magneto-inertial fusion This technique, combines laser pulses with the magnetic pinch to compress and confine an imploding plasma cloud. The field traps heat within the core, which improves the fusion rates. A similar concept is the magnetized target fusion device, which uses a magnetic field in an external metal shell to achieve the same basic goals.

Magnetic Pinches

Z-Pinch This method sends a strong current (in the z-direction) through the plasma. The current generates a magnetic field that squeezes the plasma to fusion conditions. Pinches were the first method for man-made controlled fusion. Some examples include the Dense plasma focus and the Z machine at Sandia National Laboratories.

Theta-Pinch This method sends a current inside a plasma, in the theta direction.

Inertial Electrostatic Confinement

Fusor This method uses an electric field to heat ions to fusion conditions. The machine typically uses two spherical cages, a cathode inside the anode, inside a vacuum. These machines are not considered a viable approach to net power due to their high conduction and radiation losses. They are simple enough to build that amateurs have fused atoms using them.

Polywell This designs attempts to combine magnetic confinement with electrostatic fields, to avoid the conduction losses generated by the cage. This research, however, is immature and under developed.

Other

Magnetized target fusionThis method confines hot plasma using a magnetic field and squeezes it using inertia. Examples include LANL FRX-L machine and General Fusion device.

Uncontrolled Fusion has been initiated by man, using uncontrolled fission explosions. Early proposals for fusion power included using bombs to initiate reactions.

Beam fusion A beam of high energy particles can be fired at another beam or target and fusion will occur. This was used in the 1970s and 80's to study the cross sections of high energy fusion reactions.

Bubble fusion This was a supposed fusion reaction that was supposed to occur inside extraordinarily large collapsing gas bubbles, created during acoustic liquid cavitation.

Cold fusion This is a hypothetical type of nuclear reaction that would occur at, or near, room temperature. Cold fusion has gained a reputation as Pathological science.

Muon-catalyzed fusion Muons allow atoms to get much closer and thus reduce the kinetic energy required to initiate fusion. Muons require more energy to produce than can be obtained from muon-catalysed fusion, making this approach impractical for the generation of power

Fuels

By firing particle beams at targets, many fusion reactions have been tested, while the fuels considered for power have all been light elements like the isotopes of hydrogen—deuterium and tritium. Other reactions like the deuterium and Helium^3 reaction or the Helium^3 and Helium^3 reactions, would require a supply of Helium^3. This can either come from other nuclear reactions or from extraterrestrial sources. Finally, researchers hope to do the p-11B reaction, because it does not directly produce neutrons, though side reactions can.

Deuterium, Tritium

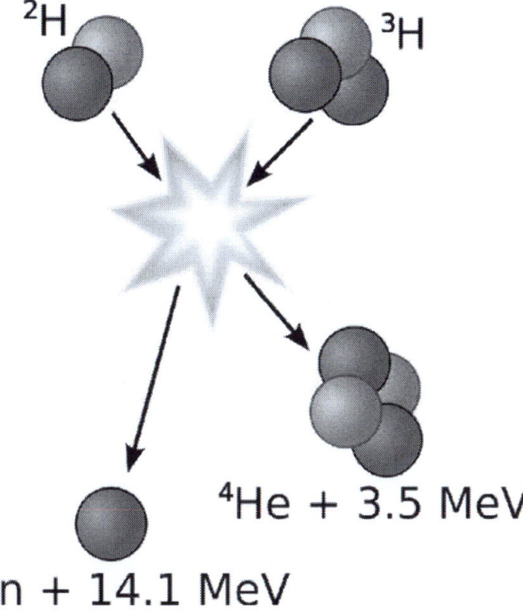

Figure: *Diagram of the D-T reaction*

The easiest nuclear reaction, at the lowest energy, is:

21D + 31T → 42He + 10n

This reaction is common in research, industrial and military applications, usually as a convenient source of neutrons. Deuterium is a naturally occurring isotope of hydrogen and is commonly available. The large mass ratio of the hydrogen isotopes makes their separation easy compared to the difficult uranium enrichment process. Tritium is a natural isotope of hydrogen, but due to its short half-life of 12.32 years, is hard to find, store, produce, and is expensive. Consequently, the deuterium-tritium fuel cycle requires the breeding of tritium from lithium using one of the following reactions:

10n + 63Li → 31T + 42He

10n + 73Li → 31T + 42He + 10n

The reactant neutron is supplied by the D-T fusion reaction shown above, and the one that has the greatest yield of energy. The reaction with ^{6}Li is exothermic, providing a small energy gain for the reactor. The reaction with ^{7}Li is endothermic but does not consume the neutron. At least some ^{7}Li reactions are required to replace the neutrons lost to absorption by other elements. Most reactor designs use the naturally occurring mix of lithium isotopes.

Several Drawbacks are Commonly Attributed to D-T Fusion Power

1. It produces substantial amounts of neutrons that result in the neutron activation of the reactor materials.
2. Only about 20% of the fusion energy yield appears in the form of charged particles with the remainder carried off by neutrons, which limits the extent to which direct energy conversion techniques might be applied.
3. It requires the handling of the radioisotope tritium. Similar to hydrogen, tritium is difficult to contain and may leak from reactors in some quantity. Some estimates suggest that this would represent a fairly large environmental release of radioactivity.

The neutron flux expected in a commercial D-T fusion reactor is about 100 times that of current fission power reactors, posing problems for material design. After a series of D-T tests at JET, the vacuum vessel was sufficiently radioactive that remote handling was required for the year following the tests.

In a production setting, the neutrons would be used to react with lithium in order to create more tritium. This also deposits the energy of the neutrons in the lithium, which would then be transferred to drive electrical production. The lithium neutron absorption reaction protects the outer portions of the reactor from the neutron flux. Newer designs, the advanced tokamak in particular, also use lithium inside the reactor core as a key element of the design. The plasma interacts directly with the lithium, preventing a problem known as "recycling". The advantage of this design was demonstrated in the Lithium Tokamak Experiment.

Deuterium, Deuterium

This fuel is commonly used by amateurs who fuse. This is second easiest fusion reaction, fusing of deuterium with itself. This reaction has two branches that occur with nearly equal probability:

$$D + D \rightarrow T + {}^{1}H$$

$$D + D \rightarrow {}^{3}He + n$$

This reaction is also common in research. The optimum energy to initiate this reaction is 15 keV, only slightly higher than the optimum for the D-T reaction. The first branch does not produce neutrons, but it does produce tritium, so that a D-D reactor will not be completely tritium-free, even though it does not require an input of tritium or lithium. Unless the tritons can be quickly removed, most of the tritium produced would be burned before leaving the reactor, which would reduce the handling of tritium, but would produce more neutrons, some of which are very energetic.

The neutron from the second branch has an energy of only 2.45 MeV (0.393 pJ), whereas the neutron from the D-T reaction has an energy of 14.1 MeV (2.26 pJ), resulting in a wider range of isotope production and material damage. When the tritons are removed quickly while allowing the ^{3}He to react, the fuel cycle is called "tritium suppressed fusion" The removed tritium decays to ^{3}He with a 12.5 year half life. By recycling the ^{3}He produced from the decay of tritium back into the fusion reactor, the fusion reactor does not require materials resistant to fast 14.1 MeV (2.26 pJ) neutrons.

Assuming complete tritium burn-up, the reduction in the fraction of fusion energy carried by neutrons would be only about 18%, so that the primary advantage of the D-D fuel cycle is that tritium breeding would not be required. Other advantages are independence from scarce lithium resources and a somewhat softer neutron spectrum. The disadvantage of D-D compared to D-T is that the energy confinement time (at a given pressure) must be 30 times longer and the power produced (at a given pressure and volume) would be 68 times less .

Assuming complete removal of tritium and recycling of ^{3}He, only 6% of the fusion energy is carried by neutrons. The tritium-suppressed D-D fusion requires an energy confinement that is 10 times longer compared to D-T and a plasma temperature that is twice as high.

Deuterium, Helium 3

A second-generation approach to controlled fusion power involves combining helium-3 (^{3}He) and deuterium (^{2}H):

$$D + {}^{3}He \rightarrow {}^{4}He + {}^{1}H$$

This reaction produces a helium-4 nucleus (^{4}He) and a high-energy proton. As with the p-^{11}B aneutronic fusion fuel cycle, most of the reaction energy is released as charged particles, reducing activation of the reactor housing and potentially allowing more efficient energy harvesting (via any of several speculative technologies). In practice, D-D side reactions produce a significant number of neutrons, resulting in p-^{11}B being the preferred cycle for aneutronic fusion.

Proton, Boron 11

If aneutronic fusion is the goal, then the most promising candidate may be the Hydrogen-1 (proton)/boron reaction, which releases alpha (helium) particles, but does not rely on neutron scattering for energy transfer.

$$^{1}H + ^{11}B \rightarrow 3\ ^{4}He$$

Under reasonable assumptions, side reactions will result in about 0.1% of the fusion power being carried by neutrons. At 123 keV, the optimum temperature for this reaction is nearly ten times higher than that for the pure hydrogen reactions, the energy confinement must be 500 times better than that required for the D-T reaction, and the power density will be 2500 times lower than for D-T.

Since the confinement properties of conventional approaches to fusion such as the tokamak and laser pellet fusion are marginal, most proposals for aneutronic fusion are based on radically different confinement concepts, such as the Polywell and the Dense Plasma Focus. Results have been extremely promising:

"In the October 2013 edition of Nature Communications, a research team led by Christine Labaune at Ecole Polytechnique in Palaiseau, France, reported a new record fusion rate: an estimated 80 million fusion reactions during the 1.5 nanoseconds that the laser fired, which is at least 100 times more than any previous proton-boron experiment."

Heating

Gas must be first heated to form a plasma. This then needs to be hot enough to start fusion reactions. A number of heating schemes have been explored:

Radiofrequency Heating A radio wave is applied to the plasma, causing it to oscillate.

Electrostatic Heating An electric field can do work on a charged ions or electrons, heating them.

Neutral Beam Injection Gas is heated and injected into the fusion device. It may be heated using an electric field and then neutralized. After injection, it collides with particles the imparting energy.

Confinement

Unconfined: The first human-made, large-scale fusion reaction was the test of the hydrogen bomb, Ivy Mike, in 1952. As part of the PACER project, it was once proposed to use hydrogen bombs as a source of power by detonating them in underground caverns and then generating electricity from the heat produced, but such a power plant is unlikely ever to be constructed.

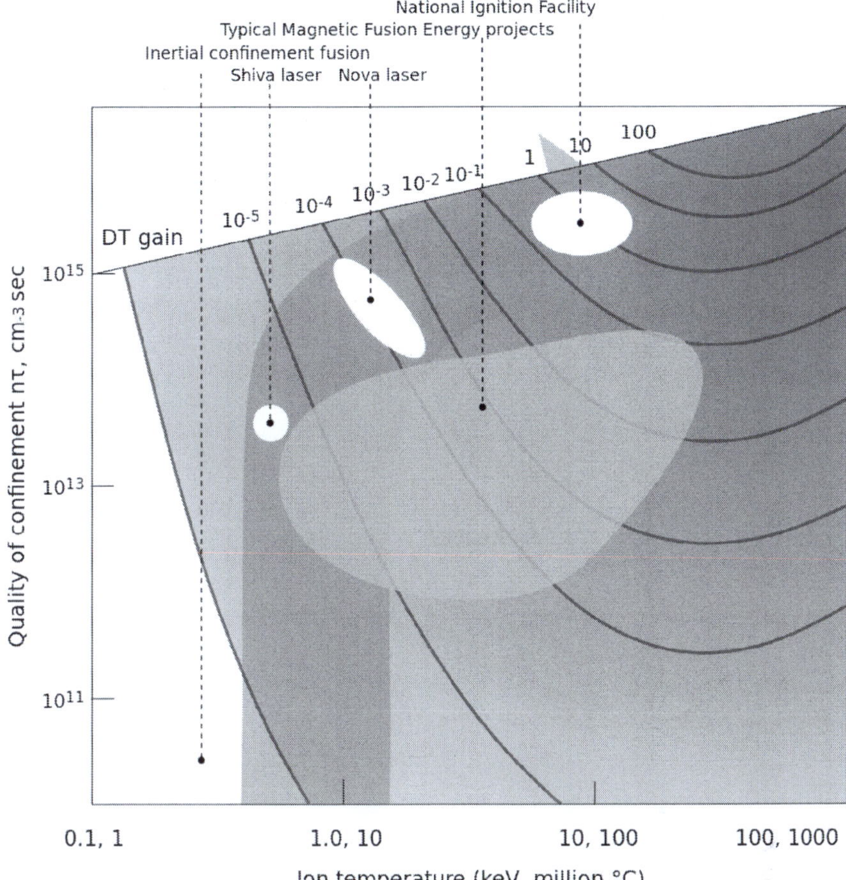

Figure: *Parameter space occupied by inertial fusion energy and magnetic fusion energy devices as of the mid* ***1990s****. The regime allowing thermonuclear ignition with high gain lies near the upper right corner of the plot.*

General Confinement Principles

Confinement refers to all the conditions necessary to keep a plasma dense and hot long enough to undergo fusion:

- Equilibrium: The forces acting on the plasma must be balanced so that it will not rapidly disassemble. The exception, of course, is inertial confinement, where the relevant physics must occur faster than the disassembly time.
- Stability: The plasma must be so constructed that small deviations are restored to the initial state, otherwise some unavoidable disturbance will occur and grow exponentially until the plasma is destroyed.

- Transport: The loss of particles and heat in all channels must be sufficiently slow. The word "confinement" is often used in the restricted sense of "energy confinement".

Controlled fusion refers to the continuous power production from fusion, or at least the use of explosions that are so small that they do not destroy a significant portion of the machine that produces them. To produce self-sustaining fusion, the energy released by the reaction (or at least a fraction of it) must be used to heat new reactant nuclei and keep them hot long enough that they also undergo fusion reactions. Retaining the energy is called energy confinement and may be accomplished in a number of ways, Material, Gravitational, Electrostatic, Inertial, and Magnetic Confinement.

Inertial Confinement

Inertial confinement is the use of rapidly imploding shell to heat and confine plasma. The shell is imploded using a direct laser blast (direct drive) or a secondary x-ray blast (indirect drive). Theoretically, fusion using lasers would be done using tiny pellets of fuel that explode several times a second. To induce the explosion, the pellet must be compressed to about 30 times solid density with energetic beams. If direct drive is used - the beams are focused directly on the pellet - it can in principle be very efficient, but in practice is difficult to obtain the needed uniformity. The alternative approach, indirect drive, uses beams to heat a shell, and then the shell radiates x-rays, which then implode the pellet. The beams are commonly laser beams, but heavy and light ion beams and electron beams have all been investigated.

Inertial confinement produces plasmas with very high densities and temperatures making it suitable for weapons research, X-ray generation and perhaps in the distant future, spaceflight . ICF implosions require fuel pellets with close to a perfect shape in order to generate an symmetrical inward shock wave and to produce the high-density plasma. These are known as targets and, building them has presented its own technical challenges.

A recent development in ICF research is fast ignition. This is the use of two laser systems to heat a compressed targets. A conventional laser system compresses the pellet, after which a second ultrashort laser pulse heats the compressed plasma. This burst has many petawatts of power. Fast ignition implodes the pellet at exactly the moment of greatest density. Research into fast ignition has been carried out at the OMEGA EP petawatt and OMEGA lasers at the University of Rochester and at the GEKKO XII laser at the institute for laser engineering in Osaka Japan. If fruitful, it may have the effect of greatly reducing the cost of a laser fusion based power source.

Magnetic Confinement

At the temperatures required for fusion, the fuel is heated to a plasma state. In this state it has a very good electrical conductivity. This opens the possibility of confining the plasma with magnetic fields, an idea known as magnetic confinement. This puts a Lorentz force on the plasma. The force works perpendicular to the magnetic fields, so one problem in magnetic confinement is preventing the plasma from leaking out the ends of the field lines. These ends are known as magnetic cusps. There are basically two solutions.

The first is to use the magnetic mirror effect. If a particle follows the field line and enters a region of higher field strength, the particles can be reflected. There are several devices that try to use this effect. The most famous was the magnetic mirror machines, which was a series of large, expensive devices built at the Lawrence Livermore National Laboratory from the 1960s to mid 1980s. Some other examples include the magnetic bottles and Biconic cusp. Because the mirror machines were straight, they had some advantages over a ring shape. First, mirrors would easier to construct and maintain and second direct conversion energy capture, was easier to implement. As the confinement achieved in experiments was poor, this approach has been essentially abandoned.

The second possibility to prevent end losses is to bend the field lines back on themselves, either in circles or more commonly in nested toroidal surfaces. The most highly developed system of this type is the *tokamak*, with the *stellarator* being next most advanced, followed by the Reversed field pinch. Compact toroids, especially the *Field-Reversed Configuration* and the spheromak, attempt to combine the advantages of toroidal magnetic surfaces with those of a simply connected (non-toroidal) machine, resulting in a mechanically simpler and smaller confinement area. Compact toroids still have some enthusiastic supporters but are not backed as readily by the majority of the fusion community.

Electrostatic Confinement

There are also electrostatic confinement fusion devices. These devices heat and confine ions using electrostatic fields. The best known is the Fusor. This device has an cathode inside an anode wire cage.

Positive ions fly towards the negative inner cage, and are heated by the electric field in the process. If they miss the inner cage they can collide and fuse. Ions typically hit the cathode, however, creating prohibitory high conduction losses. Also, fusion rates in fusors are very low due to competing physical effects, such as energy loss in the form of light radiation. Designs have been proposed to avoid the problems associated with the cage, by generating the field using a non-neutral cloud. These include a plasma oscillating device, a penning trap and the polywell. The technology is relatively immature, however, and many scientific and engineering questions remain.

Plant Design

Nuclear Island: A fusion power plant may be designed with a nuclear island and the balance of plant. This is common in typical fission reactors. The nuclear island has a plasma chamber with an associated vacuum system, surrounded by plasma-facing components (first wall and divertor) maintaining the vacuum and absorbing the heat coming from the plasma. If magnetic confinement is used, a magnet system made from superconducting magnets will be needed, as well as systems for heating and refuelling the plasma. If inertial confinement is used, it will require a driver (laser or accelerator) and a focusing system, as well as place to manufacture and position the target. The balance of plant converts heat into electricity via steam turbines.

Energy Extraction

Steam turbines It has been proposed that steam turbines be used to convert the heat from the fusion chamber into electricity. The heat is transferred into a working fluid that turns into steam, driving electric generators.

Neutron blankets Deuterium and tritium fusion generates neutrons. This varies by technique (NIF has a record of 3E14 neutrons per second while a typical fusor produces 1E5 - 1E9 neutrons per second). It has been proposed to use these neutrons as a way to regenerate spent fission fuel or as a way to breed tritium from a liquid lithium blanket.

Direct conversion This is a method where the kinetic energy of a particle is converted into voltage. It was first suggested by Richard F. Post in conjunction with magnetic mirrors, in the late sixties. It has also been suggested for Field-Reversed Configurations. The process takes the plasma, expands it, and converts a large fraction of the random energy of the fusion products into directed motion. The particles are then collected on electrodes at various large electrical potentials. This method has demonstrated an experimental efficiency of 48 percent.

Diagnostics

Thomson Scattering Certain wavelengths of light will scatter off a plasma. This light can be detected and used to reconstruct the plasmas' behaviour. This technique can be used to find its' density and temperature. It is common in Inertial confinement fusion, Tokamaks and fusors. In ICF systems, this can be done by firing a second beam into a gold foil adjacent to the target. This makes x-rays that scatter or traverse the plasma. In Tokamaks, this can be done using mirrors and detectors to reflect light across a plane (two dimensions) or in a line (one dimension).

Langmuir probe This is a metal object placed in a plasma. A potential is applied to it, giving it a positive or negative voltage against the surrounding plasma. The metal collects charged particles, drawing a current. As the voltage changes, the current changes. This makes a IV Curve. The IV-curve can be used to determine the local plasma density, potential and temperature.

Geiger counter Deuterium or tritium fusion produces neutrons. Geiger counters record the rate of neutron production, so they are an essential tool for demonstrating success.

Safety and the Environment

Accident Potential: There is no possibility of a catastrophic accident in a fusion reactor resulting in major release of radioactivity to the environment or injury to non-staff, unlike modern fission reactors. The primary reason is that the requirements for nuclear fusion differ greatly from nuclear fission: fusion requires extremely precise and controlled temperature, pressure, and magnetic field parameters for any net energy to be produced, and a far smaller amount of fuel. If the reactor suffered damage or lost even a small degree of required control, fusion reactions and heat generation would rapidly cease.

Therefore fusion reactors are considered extremely safe in this sense, making them favourable over fission reactors, which, in contrast, continue to generate heat through beta-decay for several months after reactor shut-down, meaning that melting of fuel rods is possible even after the reactor has been stopped due to continued accumulation of heat.

There is also no risk of a runaway reaction in a fusion reactor. The plasma is burnt at optimal conditions, and any significant change will render it unable to react or to produce excess heat. In fusion reactors the reaction process is so delicate that this level of safety is inherent; no elaborate failsafe mechanism is required.

Although the plasma in a fusion power plant will have a volume of 1000 cubic meters or more, the density of the plasma is extremely low, and the total amount of fusion fuel in the vessel is very small, typically a few grams. If the fuel supply is closed, the reaction stops within seconds. In comparison, a fission reactor is typically loaded with enough fuel for several years, and no additional fuel is necessary to keep the reaction going.

In the magnetic approach, strong fields are developed in coils that are held in place mechanically by the reactor structure. Failure of this structure could release this tension and allow the magnet to "explode" outward. The severity of this event would be similar to any other industrial accident or an MRI machine quench/explosion, and could be effectively stopped with a containment building similar to those used in existing (fission) nuclear generators. The laser-driven inertial approach is generally lower-stress. Although failure of the reaction chamber is possible, simply stopping fuel delivery would prevent any sort of catastrophic failure.

Most reactor designs rely on the use of liquid lithium as both a coolant and a method for converting stray neutrons from the reaction into tritium, which is fed back into the reactor as fuel. Lithium is highly flammable, and in the case of a fire it is possible that the lithium stored on-site could be burned up and escape. In this case, the tritium contents of the lithium would be released into the atmosphere, posing a radiation risk. Calculations suggest that at about 1 kg the total amount of tritium and other radioactive gases in a typical power plant would be so small that they would have diluted to legally acceptable limits by the time they blew as far as the plant's perimeter fence.

The likelihood of *small industrial* accidents including the local release of radioactivity and injury to staff cannot be estimated yet. These would include accidental releases of lithium, tritium, or mishandling of decommissioned radioactive components of the reactor itself.

Effluents During Normal Operation

The natural product of the fusion reaction is a small amount of helium, which is completely harmless to life. Of more concern is tritium, which, like other isotopes of hydrogen, is difficult to retain completely. During normal operation, some amount of tritium will be continually released. There would be no acute danger, but the cumulative effect on the world's population from a fusion economy could be a matter of concern.

Although tritium is volatile and biologically active the health risk posed by a release is much lower than that of most radioactive contaminants, due to tritium's short half-life (12.32 years), very low decay energy (~14.95 keV), and the fact that it does not bioaccumulate (instead being cycled out of the body as water, with a biological half-life of 7 to 14 days). Current ITER designs are investigating total containment facilities for any tritium.

Waste Management

The large flux of high-energy neutrons in a reactor will make the structural materials radioactive. The radioactive inventory at shut-down may be comparable to that of a fission reactor, but there are important differences.

The half-life of the radioisotopes produced by fusion tends to be less than those from fission, so that the inventory decreases more rapidly. Unlike fission reactors, whose waste remains radioactive for thousands of years, most of the radioactive material in a fusion reactor would be the reactor core itself, which would be dangerous for about 50 years, and low-level waste another 100. Although this waste will be considerably more radioactive during those 50 years than fission waste, the very short half-life makes the process very attractive, as the waste management is fairly straightforward. By 500 years the material would have the same radiotoxidity as coal ash.

Additionally, the choice of materials used in a fusion reactor is less constrained than in a fission design, where many materials are required for their specific neutron cross-sections. This allows a fusion reactor to be designed using materials that are selected specifically to be "low activation", materials that do not easily become radioactive. Vanadium, for example, would become much less radioactive than stainless steel. Carbon fibre materials are also low-activation, as well as being strong and light, and are a promising area of study for laser-inertial reactors where a magnetic field is not required.

In general terms, fusion reactors would create far less radioactive material than a fission reactor, the material it would create is less damaging biologically, and the radioactivity "burns off" within a time period that is well within existing engineering capabilities for safe long-term waste storage.

Nuclear Proliferation

Although fusion power uses nuclear technology, the overlap with nuclear weapons would be limited. A huge amount of tritium could be produced by a fusion power plant. Tritium is used in the trigger of hydrogen bombs and in a modern boosted fission weapon. But tritium can be also produced by nuclear fission. The energetic neutrons from a fusion reactor could be used to breed weapons-grade plutonium or uranium for an atomic bomb (for example by transmutation of U^{238} to Pu^{239}, or Th^{232} to U^{233}).

A study conducted 2011 assessed the risk of three scenarios:

- Use in small-scale fusion plant: Due to much higher power consumption, heat dissipation and a more unique design compared to enrichment gas centrifuges this choice would be much easier to detect and therefore implausible.
- Modifications to produce weapon-usable material in a commercial facility: The production potential is significant. But no fertile or fissile substances necessary for the production of weapon-usable materials needs to be present at a civil fusion system at all. If not shielded, a detection of these materials can be done by their characteristic gamma radiation. The underlying redesign could be detected by regular design information verifications. In the (technically more feasible) case of solid breeder blanket modules, it would be necessary for incoming components to be inspected for the presence of fertile material, otherwise plutonium for several weapons could be produced each year.
- Prioritizing a fast production of weapon-grade material regardless of secrecy: The fastest way to produce weapon usable material was seen in modifying a prior civil fusion power plant. Unlike in some nuclear power plants, there is no weapon compatible material during civil use. Even without the need for covert action this modification would still take about 2 months to start the production and at least an additional week to generate a significant amount for weapon production. This was seen as enough time to detect a military use and to react with diplomatic or military means. To stop the production, a military destruction of inevitable parts of the facility leaving out the reactor itself would be sufficient. This, together with the intrinsic safety of fusion power would only bear a low risk of radioactive contamination.

Another study concludes that "[..]large fusion reactors – even if not designed for fissile material breeding – could easily produce several hundred kg Pu per year with high weapon quality and very low source material requirements." It was emphasized that the implementation of features for intrinsic proliferation resistance might only be possible at this phase of research and development. The theoretical and computational tools needed for hydrogen bomb design are closely related to those needed for inertial confinement fusion, but have very little in common with the more scientifically developed magnetic confinement fusion.

As a Sustainable Energy Source

Large-scale reactors using neutronic fuels (e.g. ITER) and thermal power production (turbine based) are most comparable to fission power from an engineering and economics viewpoint. Both fission and fusion power plants involve a relatively compact heat source powering a conventional steam turbine-based power plant, while producing enough neutron radiation to make activation of the plant materials problematic. The main distinction is that fusion power produces no high-level radioactive waste (though activated plant materials still need to be disposed of). There are some power plant ideas that may significantly lower the cost or size of such plants; however, research in these areas is nowhere near as advanced as in tokamaks.

Fusion power commonly proposes the use of deuterium, an isotope of hydrogen, as fuel and in many current designs also use lithium. Assuming a fusion energy output equal to the 1995 global power output of about 100 EJ/yr (= 1×10^{20} J/yr) and that this does not increase in the future, which is unlikely, then the known current lithium reserves would last 3000 years. Lithium from sea water would last 60 million years, however, and a more complicated fusion process using only deuterium from sea water would have fuel for 150 billion years. To put this in context, 150 billion years is close to 30 times the remaining lifespan of the sun, and more than 10 times the estimated age of the universe.

Economics

While fusion power is still in early stages of development, substantial sums have been and continue to be invested in research. In the EU almost €10 billion was spent on fusion research up to the end of the 1990s, and the new ITER reactor alone is budgeted at €10 billion.

It is estimated that up to the point of possible implementation of electricity generation by nuclear fusion, R&D will need further promotion totalling around €60–80 billion over a period of 50 years or so (of which €20–30 billion within the EU) based on a report from 2002. Nuclear fusion research receives €750 million (excluding ITER funding) from the European Union, compared with €810 million for sustainable energy research, putting research into fusion power well ahead of that of any single rivaling technology. Indeed, the size of the investments and time frame of the expected results mean that fusion research is almost exclusively publicly funded, while research in other forms of energy can be done by the private sector.

Advantages

Fusion power would provide more energy for a given weight of fuel than any fuel-consuming energy source currently in use, and the fuel itself (primarily deuterium) exists abundantly in the Earth's ocean: about 1 in 6500 hydrogen atoms in seawater is deuterium. Although this may seem a low proportion (about 0.015%), because nuclear fusion reactions are so much more energetic than chemical combustion and seawater is easier to access and more plentiful than fossil fuels, fusion could potentially supply the world's energy needs for millions of years.

Despite being technically non-renewable, fusion power has many of the benefits of renewable energy sources (such as being a long-term energy supply and emitting no greenhouse gases) as well as some of the benefits of the resource-limited energy sources as hydrocarbons and nuclear fission (without reprocessing). Like these currently dominant energy sources, fusion could provide very high power-generation density and uninterrupted power delivery (due to the fact that it is not dependent on the weather, unlike wind and solar power).

Another aspect of fusion energy is that the cost of production does not suffer from diseconomies of scale. The cost of water and wind energy, for example, goes up as the optimal locations are developed first, while further generators must be sited in less ideal conditions. With fusion energy the production cost will not increase much even if large numbers of plants are built, because the raw resource (seawater) is abundant and widespread.

Some problems that are expected to be an issue in this century, such as fresh water shortages, can alternatively be regarded as problems of energy supply. For example, in desalination plants, seawater can be purified through distillation or reverse osmosis. Nonetheless, these processes are energy intensive. Even if the first fusion plants are not competitive with alternative sources, fusion could still become competitive if large-scale desalination requires more power than the alternatives are able to provide.

A scenario has been presented of the effect of the commercialization of fusion power on the future of human civilization. ITER and later Demo are envisioned to bring online the first commercial nuclear fusion energy reactor by 2050. Using this as the starting point and the history of the uptake of nuclear fission reactors as a guide, the scenario depicts a rapid take up of nuclear fusion energy starting after the middle of this century.

Fusion power could be used in interstellar space, where solar energy is not available.

Chapter 5

Generation and Transmission of Electricity, Energy Storage and Fuel Cell

Electricity Generation

Electricity generation is the process of generating electric power from other sources of primary energy. The fundamental principles of electricity generation were discovered during the 1820s and early 1830s by the British scientist Michael Faraday.

Figure: *Turbo generator*

His basic method is still used today: electricity is generated by the movement of a loop of wire, or disc of copper between the poles of a magnet. For electric utilities, it is the first process in the delivery of electricity to consumers. The other processes, electricity transmission, distribution, and electrical power storage and recovery using pumped-storage methods are normally carried out by the electric power industry. Electricity is most often generated at a power station by electromechanical generators, primarily driven by heat engines fuelled by chemical combustion or nuclear fission but also by other means such as the kinetic energy of flowing water and wind. Other energy sources include solar photovoltaics and geothermal power.

Methods of Generating Electricity

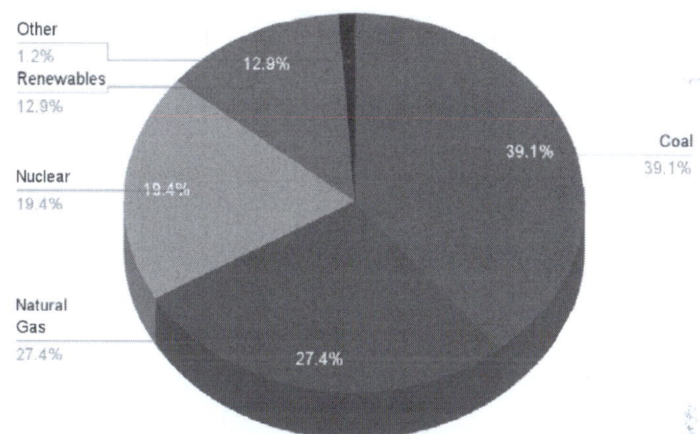

Figure: U.S. 2013 Electricity Generation By Type.

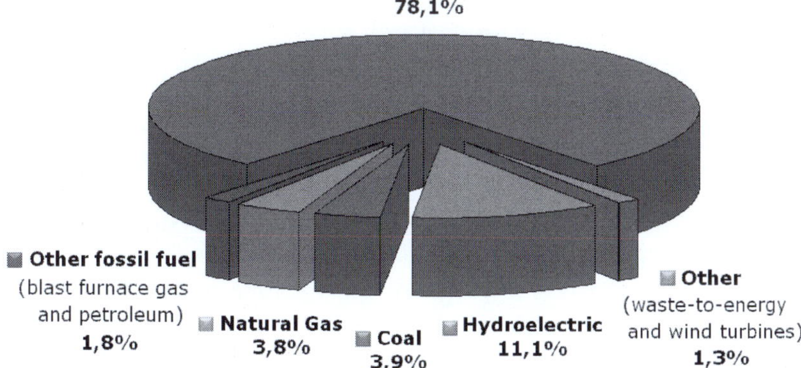

Figure: *Sources of electricity in France in 2006; nuclear power was the main source.*

There are seven fundamental methods of directly transforming other forms of energy into electrical energy:

- Static electricity, from the physical separation and transport of charge (examples: triboelectric effect and lightning)
- Electromagnetic induction, where an electrical generator, dynamo or alternator transforms kinetic energy (energy of motion) into electricity. This is the most used form for generating electricity and is based on Faraday's law. It can be experimented by simply rotating a magnet within closed loops of a conducting material (e.g. copper wire)
- Electrochemistry, the direct transformation of chemical energy into electricity, as in a battery, fuel cell or nerve impulse
- Photovoltaic effect, the transformation of light into electrical energy, as in solar cells
- Thermoelectric effect, the direct conversion of temperature differences to electricity, as in thermocouples, thermopiles, and thermionic converters.
- Piezoelectric effect, from the mechanical strain of electrically anisotropic molecules or crystals. Researchers at the US Department of Energy's Lawrence Berkeley National Laboratory (Berkeley Lab) have developed a piezoelectric generator sufficient to operate a liquid crystal display using thin films of M13 bacteriophage.
- Nuclear transformation, the creation and acceleration of charged particles (examples: betavoltaics or alpha particle emission)

Static electricity was the first form discovered and investigated, and the electrostatic generator is still used even in modern devices such as the Van de Graaff generator and MHD generators. Charge carriers are separated and physically transported to a position of increased electric potential. Almost all commercial electrical generation is done using electromagnetic induction, in which mechanical energy forces an electrical generator to rotate. There are many different methods of developing the mechanical energy, including heat engines, hydro, wind and tidal power. The direct conversion of nuclear potential energy to electricity by beta decay is used only on a small scale. In a full-size nuclear power plant, the heat of a nuclear reaction is used to run a heat engine. This drives a generator, which converts mechanical energy into electricity by magnetic induction. Most electric generation is driven by heat engines. The combustion of fossil fuels supplies most of the heat to these engines, with a significant fraction from nuclear fission and some from renewable sources. The modern steam turbine (invented by Sir Charles Parsons in 1884) currently generates about 80% of the electric power in the world using a variety of heat sources.

Economics of Generation and Production of Electricity

The selection of electricity production modes and their economic viability varies in accordance with demand and region. The economics vary considerably around the world, resulting in widespread selling prices, e.g. the price in Venezuela is 3 cents per kWh while in Denmark it is 40 cents per kWh. Hydroelectric plants, nuclear power plants, thermal power plants and renewable sources have their own pros and cons, and selection is based upon the local power requirement and the fluctuations in demand. All power grids have varying loads on them but the daily minimum is the base load, supplied by plants which run continuously. Nuclear, coal, oil and gas plants can supply base load.

Thermal energy is economical in areas of high industrial density, as the high demand cannot be met by renewable sources. The effect of localized pollution is also minimized as industries are usually located away from residential areas. These plants can also withstand variation in load and consumption by adding more units or temporarily decreasing the production of some units. Nuclear power plants can produce a huge amount of power from a single unit. However, recent disasters in Japan have raised concerns over the safety of nuclear power, and the capital cost of nuclear plants is very high. Hydroelectric power plants are located in areas where the potential energy from falling water can be harnessed for moving turbines and the generation of power. It is not an economically viable source of production where the load varies too much during the annual production cycle and the ability to store the flow of water is limited.

Renewable sources other than hydroelectricity (solar power, wind energy, tidal power, etc.) due to advancements in technology, and with mass production, their cost of production has come down and the energy is now in many cases cost-comparative with fossil fuels. Many governments around the world provide subsidies to offset the higher cost of any new power production, and to make the installation of renewable energy systems economically feasible. However, their use is frequently limited by their intermittent nature. If natural gas prices are below $3 per million British thermal units, generating electricity from natural gas is cheaper than generating power by burning coal.

Production

The production of electricity in 2009 was 20,053TWh. Sources of electricity were fossil fuels 67%, renewable energy 16% (mainly hydroelectric, wind, solar and biomass), and nuclear power 13%, and other sources were 3%. The majority of fossil fuel usage for the generation of electricity was coal and gas. Oil was 5.5%, as it is the most expensive common commodity used to produce electrical energy. Ninety-two percent of renewable energy was hydroelectric followed by wind at 6% and geothermal at 1.8%. Solar photovoltaic was 0.06%, and solar thermal was 0.004%. Data are from OECD 2011-12 Factbook (2009 data).

Source of Electricity (World total year 2008)							
-	*Coal*	*Oil*	*Natural Gas*	*Nuclear*	*Renewables*	*other*	*Total*
Average electric power (TWh/year)	8,263	1,111	4,301	2,731	3,288	568	20,261
Average electric power (GW)	942.6	126.7	490.7	311.6	375.1	64.8	2311.4
Proportion	41%	5%	21%	13%	16%	3%	100%

data source IEA/OECD

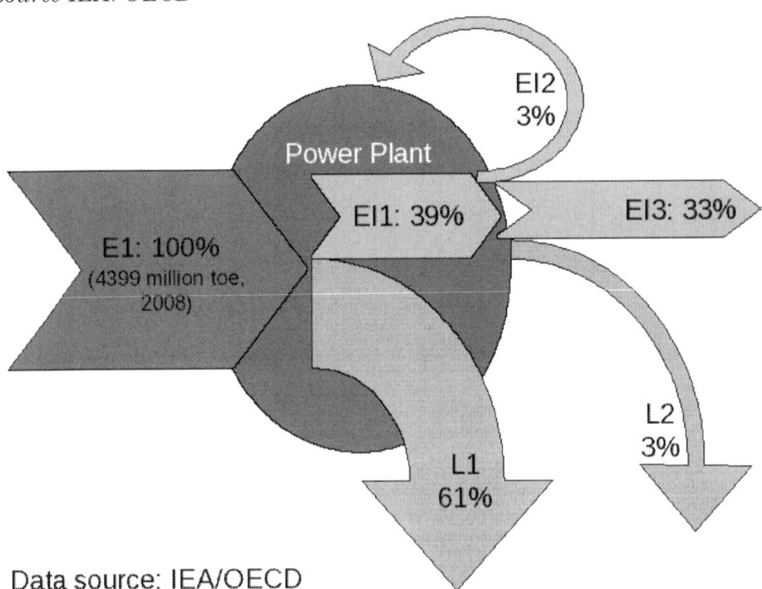

Figure: *Energy Flow of Power Plant*

Total energy consumed at all power plants for the generation of electricity was 4,398,768 ktoe (kilo ton of oil equivalent) which was 36% of the total for primary energy sources (TPES) of 2008.

Electricity output (gross) was 1,735,579 ktoe (20,185 TWh), efficiency was 39%, and the balance of 61% was generated heat. A small part (145,141 ktoe, which was 3% of the input total) of the heat was utilised at co-generation heat and power plants. The in-house consumption of electricity and power transmission losses were 289,681 ktoe. The amount supplied to the final consumer was 1,445,285 ktoe (16,430 TWh) which was 33% of the total energy consumed at power plants and heat and power co-generation (CHP) plants.

Power Transmission Line?

A power transmission line is simply a conductor cable that is used to transport electricity from one place to another. In the industrial vernacular, this usually means from the power generating plant to the substation. Lines that take the power from the substation to the individual customers are known as power distribution lines. Despite this distinction, nearly any power line is known simply as a power transmission line to the layperson.

In most cases, a power transmission line consists of multiple conductive wires approximately one inch (25.4 mm) in diameter. There are usually three of these wires. Along with these wires is a protective wire at the top known as a shield wire. The job of the shield wire is to stop lightning as much as possible. This not only helps prevent lines from failing, but also helps to prevent power surges that can accompany electrical storms and damage electronic equipment in a home.

The line that runs from the power plant to a substation is often known as a high power transmission line. This line will carry the most voltage. A high-voltage electric power transmission line can carry as much as 765 kilovolts. In contrast, the line that runs into a home carries as much as 24 kilovolts. The advantage of high-power lines is that they offer the opportunity to transmit electricity hundreds of miles without much loss.

The voltage is only one feature of these lines. Though some are located underground, most power transmission lines are located above ground, where they are elevated to such a point that interference with most human activities is rare. The above-ground power transmission line offers several advantages. They are cheaper to install, easier to maintain, and easier to troubleshoot than underground lines.

While these advantages often compel companies to choose an above-ground option, there are disadvantages as well. During ice storms or heavy wind events, power lines may become stressed to the point that poles supporting the lines collapse. Further, many complain about the aesthetics of these lines, arguing they often run through natural areas and create an unsightly view.

The high voltage power transmission line has long been the focus of study for those concerned about the electromagnetic fields generated around these lines. Many feared they may raise the risk of certain types of cancer, especially childhood leukemia. The U.S. National Institute of Environmental Health Sciences suggested the connection between childhood leukemia and high power transmission lines was weak at best. Still, the concern may be enough for some people to avoid living near such lines, if that is an option.

Electrical Transformers

Electrical transformers are used to transform electrical energy. How they do so is by altering voltage, generally from high to low. Voltage is simply the measurement of electrons, how many or how strong, in the flow. Electricity can then be transported more easily and efficiently over long distances.

While power line electrical transformers are commonly recognised, there are other various types and sizes as well. They range from huge, multi-ton units like those at power plants, to intermediate, such as the type used on electric poles, and others can be quite small. Those used in equipment or appliances in your home or place of business are smaller transformers and there are also tiny ones used in items like microphones and other electronics.

Probably the most common and perhaps the most necessary use of various electrical transformers is the transportation of electricity from power plants to homes and businesses. Because power often has to travel long distances, it is transformed first into a more manageable state. It is then transformed again and again, or "stepped down," repeatedly as it gets closer to its destination.

When the power leaves the plant, it is usually of high voltage. When it reaches the substation the voltage is lowered. When it reaches a smaller transformer, the type found on top of electric poles, it is stepped down again. It is a continuous process, which repeats until the power is at a usable level.

You have likely seen the type of electrical transformers that sit on top of electric poles. These contain coils or windings that are wrapped around a core. The power travels through the coils. The more coils, the higher the voltage. On the other hand, fewer coils mean lower voltage.

Electrical transformers have changed industry. Electric power distribution is now more efficient than ever. Transformers have made it possible to transfer power near and far, in a timely, efficient, and more economical manner. Since many people do not wish to live in close proximity to a power plant, there is the added benefit of making it possible for homes and businesses that are quite a distance from power plants to obtain dependable, affordable electricity.

Electrical Grid

An electrical grid (also referred to as an electricity grid or electric grid) is an interconnected network for delivering electricity from suppliers to consumers. It consists of generating stations that produce electrical power, high-voltage transmission lines that carry power from distant sources to demand centres, and distribution lines that connect individual customers.

Power stations may be located near a fuel source, at a dam site, or to take advantage of renewable energy sources, and are often located away from heavily populated areas. They are usually quite large to take advantage of the economies of scale. The electric power which is generated is stepped up to a higher voltage-at which it connects to the transmission network.

The transmission network will move the power long distances, sometimes across international boundaries, until it reaches its wholesale customer (usually the company that owns the local distribution network).

On arrival at a substation, the power will be stepped down from a transmission level voltage to a distribution level voltage. As it exits the substation, it enters the distribution wiring. Finally, upon arrival at the service location, the power is stepped down again from the distribution voltage to the required service voltage(s).

Features

Structure of Distribution Grids: The structure, or "topology" of a grid can vary considerably. The physical layout is often forced by what land is available and its geology. The logical topology can vary depending on the constraints of budget, requirements for system reliability, and the load and generation characteristics.

The cheapest and simplest topology for a distribution or transmission grid is a *radial* structure. This is a *tree* shape where power from a large supply radiates out into progressively lower voltage lines until the destination homes and businesses are reached.

Most transmission grids require the reliability that more complex *mesh networks* provide. If one were to imagine running redundant lines between limbs/branches of a tree that could be turned in case any particular limb of the tree were severed, then this image approximates how a mesh system operates. The expense of mesh topologies restrict their application to transmission and medium voltage distribution grids. Redundancy allows line failures to occur and power is simply rerouted while workmen repair the damaged and deactivated line.

Other topologies used are *looped* systems found in Europe and *tied ring* networks.

In cities and towns of North America, the grid tends to follow the classic *radially fed* design. A substation receives its power from the transmission network, the power is stepped down with a transformer and sent to a bus from which feeders fan out in all directions across the countryside. These feeders carry three-phase power, and tend to follow the major streets near the substation. As the distance from the substation grows, the fanout continues as smaller laterals spread out to cover areas missed by the feeders. This tree-like structure grows outward from the substation, but for reliability reasons, usually contains at least one unused backup connection to a nearby substation. This connection can be enabled in case of an emergency, so that a portion of a substation's service territory can be alternatively fed by another substation.

Geography of Transmission Networks

Transmission networks are more complex with redundant pathways.

A wide area synchronous grid or "interconnection" is a group of distribution areas all operating with alternating current (AC) frequencies synchronized (so that peaks occur at the same time). This allows transmission of AC power throughout the area, connecting a large number of electricity generators and consumers and potentially enabling more efficient electricity markets and redundant generation. Interconnection maps are shown of North America (right) and Europe (below left).

In a synchronous grid all the generators run not only at the same frequency but also at the same phase, each generator maintained by a local governor that regulates the driving torque by controlling the steam supply to the turbine driving it. Generation and consumption must be balanced across the entire grid, because energy is consumed almost instantaneously as it is produced. Energy is stored in the immediate short term by the rotational kinetic energy of the generators.

A large failure in one part of the grid - unless quickly compensated for - can cause current to re-route itself to flow from the remaining generators to consumers over transmission lines of insufficient capacity, causing further failures. One downside to a widely connected grid is thus the possibility of cascading failure and widespread power outage. A central authority is usually designated to facilitate communication and develop protocols to maintain a stable grid. For example, the North American Electric Reliability Corporation gained binding powers in the United States in 2006, and has advisory powers in the applicable parts of Canada and Mexico. The U.S. government has also designated National Interest Electric Transmission Corridors, where it believes transmission bottlenecks have developed.

Some areas, for example rural communities in Alaska, do not operate on a large grid, relying instead on local diesel generators.

High-voltage direct current lines or variable frequency transformers can be used to connect two alternating current interconnection networks which are not synchronized with each other. This provides the benefit of interconnection without the need to synchronize an even wider area. For example, compare the wide area synchronous grid map of Europe (above left) with the map of HVDC lines (below right).

Redundancy and Defining "Grid"

A town is only said to have achieved grid connection when it is connected to several redundant sources, generally involving long-distance transmission.

This redundancy is limited. Existing national or regional grids simply provide the interconnection of facilities to utilise whatever redundancy is available. The exact stage of development at which the supply structure becomes a *grid* is arbitrary. Similarly, the term *national grid* is something of an anachronism in many parts of the world, as transmission cables now frequently cross national boundaries. The terms *distribution grid* for local connections and *transmission grid* for long-distance transmissions are therefore preferred, but *national grid* is often still used for the overall structure.

Interconnected Grid

Electric utilities across regions are many times interconnected to allow for a variety of advantages. First is the fact that electric utilities benefits from its nature of being large and interconnecting utilities allows for economies of scale. Second, utilities can draw power from generator reserves from a different region in order to ensure continuing, reliable power and diversify their loads. Interconnection also allows regions to have access to cheap bulk energy by receiving power from different sources. For example, one region may be producing cheap hydro power during high water seasons, but in low water seasons, another area may be producing cheaper power through wind, allowing both regions to access cheaper energy sources from one another during different times of the year. Neighbouring utilities also help others to maintain the overall system frequency and also help manage tie transfers between utility regions.

Energy Storage

Energy storage is accomplished by devices or physical media that store energy to perform useful processes at a later time. A device that stores energy is sometimes called an accumulator.

Figure: *The Llyn Stwlan dam of the Ffestiniog Pumped Storage Scheme in Wales. The lower power station has four water turbines which can generate a total of 360 MW of electricity for several hours, an example of artificial energy storage and conversion.*

Many forms of energy produce useful work, heating or cooling to meet societal needs. These energy forms include chemical energy, gravitational potential energy, electrical potential, electricity, temperature differences, latent heat, and kinetic energy. Energy storage involves converting energy from forms that are difficult to store (electricity, kinetic energy, etc.) to more conveniently or economically storable forms. Some technologies provide only short-term energy storage, and others can be very long-term such as power to gas using hydrogen or methane and the storage of heat or cold between opposing seasons in deep aquifers or bedrock. A wind-up clock stores potential energy (in this case mechanical, in the spring tension), a rechargeable battery stores readily convertible chemical energy to operate a mobile phone, and a hydroelectric dam stores energy in a reservoir as gravitational potential energy. Ice storage tanks store ice (thermal energy in the form of latent heat) at night to meet peak demand for cooling. Fossil fuels such as coal and gasoline store ancient energy derived from sunlight by organisms that later died, became buried and over time were then converted into these fuels. Even food (which is made by the same process as fossil fuels) is a form of energy stored in chemical form.

Modern Era Developments

Storing energy allows humans to balance the supply and demand of energy. Energy storage systems in commercial use today can be broadly categorized as mechanical, electrical, chemical, biological and thermal.

Storage for Electricity

Energy storage became a dominant factor in economic development with the widespread introduction of electricity. Unlike other common energy storage in prior use such as wood or coal, electricity must be used as it is being generated, or converted immediately into another form of energy such as potential, kinetic or chemical. A traditional way of storing energy on a large scale is through the use of pumped-storage hydroelectricity. Some areas of the world such as Norway, Washington and Oregon in the United States, and Wales in the United Kingdom, have used geographic features to store large quantities of water in elevated reservoirs, using excess electricity at times of low demand to pump water up into their reservoirs. The facilities then release the water which passes through turbine generators and converts the stored potential energy back to electricity when electrical demand peaks. In another example, pumped-storage hydroelectricity in Norway has an instantaneous capacity of 25–30 GW that can be expanded to 60 GW—enough to be the battery of Europe—with efforts underway in 2014 to expand its power transfer links with Germany.

Another early solution to the problem of storing energy for electrical purposes was the development of the battery as an electrochemical storage device. Batteries have previously been of limited use in electric power systems due to their relatively small capacity and high cost. However, since about the middle of the first decade of the 21st century newer battery technologies have been developed that can now provide significant utility scale load-leveling and frequency regulation capabilities. As of 2013 some of the newer battery chemistries have shown promise of being competitive with alternate energy storage methods.

Other possible large-scale methods of commercial energy storage include: flywheel, compressed air energy storage, hydrogen storage, thermal energy storage, and power to gas. Smaller scale commercial application-specific storage methods include flywheels, capacitors and supercapacitors.

Short-Term Thermal Storage, as Heat or Cold

In the 1980s, a number of manufacturers carefully researched thermal energy storage (TES) to meet the growing demand for air conditioning during peak hours. Today, several companies manufacture TES systems. The most popular form of thermal energy storage for cooling is ice storage, since it can store more energy in less space than water storage and it is also less costly than energy recovered via fuel cells or flywheels. In 2009, thermal storage used in over 3,300 buildings in over 35 countries. It works by creating ice at night when electricity is usually less costly, and then using the ice to cool the air in buildings during the hotter daytime periods.

Latent heat can also be stored in technical phase change materials (PCMs), besides ice. These can for example be encapsulated in wall and ceiling panels, to moderate room temperatures between daytime and night-time.

Interseasonal Thermal Storage, as Heat or Cold

Another class of thermal storage that has been developed since the 1970s that is now frequently employed is seasonal thermal energy storage (STES). It allows heat or cold to be used even months after it was collected from waste energy or natural sources, even in an opposing season. The thermal storage may be accomplished in contained aquifers, clusters of boreholes in geological substrates as diverse as sand or crystalline bedrock, in lined pits filled with gravel and water, or water-filled mines. An example is Alberta, *Canada's Drake Landing Solar Community,* for which 97% of the year-round heat is provided by solar-thermal collectors on the garage roofs, with a borehole thermal energy store (BTES) being the enabling technology. STES projects often have paybacks in the four-to-six year range.

Energy Storage in Chemical Fuels

Chemical fuels have become the dominant form of energy storage, both in electrical generation and energy transportation. Chemical fuels in common use are processed coal, gasoline, diesel fuel, natural gas, liquefied petroleum gas (LPG), propane, butane, ethanol and biodiesel. All of these materials are readily converted to mechanical energy and then to electrical energy using heat engines (via turbines or other internal combustion engines, or boilers or other external combustion engines) used for electrical power generation. Heat-engine-powered generators are nearly universal, ranging from small engines producing only a few kilowatts to utility-scale generators with ratings up to 800 megawatts. A key disadvantage to hydrocarbon fuels are their significant emissions of greenhouse gases that contribute to global warming, as well as other significant pollutants emitted by the dirtier fuel sources such as coal and gasoline.

Liquid hydrocarbon fuels are the most commonly used forms of energy storage for use in transportation, but because the byproducts of the reaction that utilises these liquid fuels' energy (combustion) produce greenhouse gases other energy carriers like hydrogen can be used to avoid the production of greenhouse gases.

Advanced Systems

Several advanced technologies have been investigated and are undergoing commercial development, including flywheels, which can store kinetic energy, and compressed air storage that can be pumped into underground caverns and abandoned mines to store potential energy.

Figure: *A Flybrid Kinetic Energy Recovery System flywheel. Built for use on Formula 1 racing cars, it is employed to recover and reuse kinetic energy captured during braking.*

Another advanced method used at the Solar Project in the United States and the Solar Tres Power Tower in Spain uses molten salt to store thermal energy captured from solar power and then convert it and dispatch it as electrical power when needed. The system pumps molten salt through a tower or other special conduits that are intensely heated by the sun's rays. Insulated tanks store the hot salt solution, and when needed water is then used to create steam that is fed to turbines to generate electricity.

Research is also being conducted on harnessing the quantum effects of nanoscale capacitors to create digital quantum batteries. Although this technology is still in the experimental stage, it theoretically has the potential to provide dramatic increases in energy storage capacity.

Grid Energy Storage

Grid energy storage (or large-scale energy storage) lets energy producers send excess electricity over the electricity transmission grid to temporary electricity storage sites that subsequently become energy suppliers when electricity demand is greater. Grid energy storage is particularly important in matching supply and demand over a 24 hour period of time.

A proposed variant of grid energy storage is called vehicle-to-grid energy storage system, where modern electric vehicles that are plugged into the energy grid can release the stored electrical energy in their batteries back into the grid when needed.

Renewable Energy Storage

Many renewable energy sources (most notably solar and wind) produce intermittent power. Wherever intermittent power sources reach high levels of grid penetration, energy storage becomes one option to provide reliable energy supplies. Individual energy storage projects augment electrical grids by capturing excess electrical energy during periods of low demand and storing it in other forms until needed on an electrical grid. The energy is later converted back to its electrical form and returned to the grid as needed.

Common forms of renewable energy storage include pumped-storage hydroelectricity, which has long maintained the largest total capacity of stored energy worldwide, as well as rechargeable battery systems, thermal energy storage including molten salts which can efficiently store and release very large quantities of heat energy, and compressed air energy storage. Less common, specialised forms of storage include flywheel energy storage systems, the use of cryogenic stored energy, and even superconducting magnetic coils.

Other options include recourse to peaking power plants that utilise a power-to-gas methane creation and storage process (where excess electricity is converted to hydrogen via electrolysis, combined with CO_2 (low to neutral CO_2 system) to produce methane (synthetic natural gas via the sabatier process) with stockage in the natural gas network) and smart grids with advanced energy demand management. The latter involves bringing "prices to devices", i.e. making electrical equipment and appliances able to adjust their operation to seek the lowest spot price of electricity. On a grid with a high penetration of renewables, low spot prices would correspond to times of high availability of wind and/or sunshine.

Figure: *The 150 MW Andasol solar power station is a commercial parabolic trough solar thermal power plant, located in Spain. The Andasol plant uses tanks of molten salt to store solar energy so that it can continue generating electricity even when the sun isn't shining.*

Another energy storage method is the consumption of surplus or low-cost energy (typically during night time) for conversion into resources such as hot water, cool water or ice, which is then used for heating or cooling at other times when electricity is in higher demand and at greater cost per kilowatt hour (KWh). Such thermal energy storage is often employed at end-user sites such as large buildings, and also as part of district heating, thus 'shifting' energy consumption to other times for better balancing of supply and demand.

Seasonal thermal energy storage (STES) stores heat deep in the ground via a cluster of boreholes. The Drake Landing Solar Community in Alberta, Canada has achieved a 97% solar fraction for year-round heating, with solar collectors on the garage roofs as the heat source. In

Braestrup, Denmark, the community's solar district heating system also utilises STES, at a storage temperature of 65°C (149°F). A heat pump, which is run only when there is surplus wind power available on the national grid, is used when extracting heat from the storage to raise the temperature to 80°C (176°F) for distribution.

This helps stabilize the national grid, as well as contributing to maximal use of wind power. When surplus wind generated electricity is not available, a gas-fired boiler is used. Presently, 20% of Braestrup's heat is solar, but expansion of the facility is planned to raise the fraction to 50%.

In 2011, the Bonneville Power Administration in Northwestern United States created an experimental program to absorb excess wind and hydro power generated at night or during stormy periods that are accompanied by high winds. Under computerized central control, home appliances in the region are commanded to absorb surplus energy at such times by heating ceramic bricks in special space heaters to hundreds of degrees, and by also boosting the temperature of modified hot water heater tanks. After being fully charged the highly insulated home appliances then provide home heating and hot water at later times as needed. The experimental system was created as a result of a severe 2010 storm that overproduced renewable energy in the U.S. Northwest to the extent that all conventional power sources were completely shut down, or in the case of a nuclear powerplant, reduced to its lowest possible operating level, leaving a large swath of the region running almost completely on renewable energy.

Research

Research in energy storage is being coordinated by several governments.

The German Federal government has allocated €200M (approximately US$270M) for advanced research, as well as providing a further €50M to subsidize battery storage for use with residential rooftop solar panels, according to a representative of the German Energy Storage Association.

Economic and Technical Evaluations

The economic valuation of large-scale applications (including pumped hydro storage and compressed air) must evaluate benefits including: wind curtailment avoidance, grid congestion avoidance, price arbitrage, and carbon free energy delivery. In one technical assessment by the Carnegie Mellon Electricity Industry Centre, economic goals could be met with batteries if energy storage were achievable at a capital cost of $30 to $50 per kilowatt-hour of storage capacity.

In 2014, several research and test centrees opened to evaluate energy storage technologies and efficacy. Among them in the United States was the Advanced Systems Test Laboratory at the University of Wisconsin at Madison in Wisconsin State, which partnered with the multinational conglomerate (and battery manufacturer) Johnson Controls. The laboratory was created as part of the university's newly opened Wisconsin Energy Institute. Their goals include the evaluation of state-of-the-art and next generation electric vehicle batteries, including the use of those batteries when they are connected to the electrical grid in order to supplement it during demand peaks, according to Professor Tom Jahns.

Also in 2014, the State of New York unveiled its New York Battery and Energy Storage Technology (NY-BEST) Test and Commercialization Centree at Eastman Business Park in Rochester, New York, at a cost of $23 million for its almost 1,700 m^2 laboratory. The centree, a consortium, also includes the Centree for Future Energy Systems, a collaboration between Cornell University of Ithaca, New York and the Rensselaer Polytechnic Institute in Troy, New York. NY-BEST will conduct testing, validation and independent certification of diverse forms of energy storage intended for commercial use. The centree's director stated there were currently 3,000 New Yorkers working in the energy storage industry, expected to eventually grow to 40,000 as the sector matures.

In the United Kingdom, some fourteen industry and government agencies allied themselves with seven British universities in May 2014 to create the SUPERGEN Energy Storage Hub in order to assist in the coordination of energy storage technology research and development.

Germany's Siemens AG conglomerate began commissioning a production-research plant to open in 2015 at the *Zentrum für Sonnenenergie und Wasserstoff* (ZSW, the German Centree for Solar Energy and Hydrogen Research in the State of Baden-Württemberg), a collaboration of industry plus universities in Stuttgart, Ulm and Widderstall, staffed by approximately 350 scientists, researchers, engineers, and technicians. The plant will develop new near-production manufacturing materials and processes (NPMM&P) using a computerized Supervisory Control and Data Acquisition (SCADA) system. Its goals will enable the expansion of rechargeable battery production with both increased quality and reduced manufacturing costs.

Storage Methods

Mechanical Storage: A mass of 1 kg, elevated to a height of 1,000 metres stores 9.8 kJ of gravitational energy, which is equivalent to 1 kg mass accelerated to 140 m/s. The same amount of energy is required to raise the temperature of 1 kg of water by 2.34°C.

Energy can be stored in water pumped to a higher elevation using pumped storage methods and also by moving solid matter to higher locations. Several companies such as Energy Cache and Advanced Rail Energy Storage (ARES) are working on this. Other commercial mechanical methods include compressing air and the spinning of large flywheels which converts electric energy into kinetic energy, and then back again when electrical demand peaks.

Pumped-Storage Hydroelectricity: Worldwide, pumped-storage hydroelectricity is the largest-capacity form of grid energy storage available, and, as of March 2012, the Electric Power Research Institute (EPRI) reports that PSH accounts for more than 99% of bulk storage capacity worldwide, representing around 127,000 MW. PSH reported energy efficiency varies in practice between 70% and 80%, with some claiming up to 87%.

Figure: *The Sir Adam Beck Generating Complex at Niagara Falls, Canada, which includes a large pumped storage hydroelectricity reservoir to provide an extra 174 MW of electricity during periods of peak demand.*

At times of low electrical demand, excess generation capacity is used to pump water from a lower source into a higher reservoir. When there is higher demand, water is released back into a lower reservoir (or waterway or body of water) through a turbine, generating electricity. Reversible turbine-generator assemblies act as both a pump and turbine (usually a Francis turbine design). Nearly all facilities use the height difference between two natural bodies of water or artificial reservoirs. Pure pumped-storage plants just shift the water between reservoirs, while the "pump-back" approach is a combination of pumped storage and conventional hydroelectric plants that use natural stream-flow.

Compressed Air Energy Storage: Compressed air energy storage (CAES) is a way to store energy generated at one time for use at another time using compressed air. At utility scale, energy generated during periods of low energy demand (off-peak) can be released to meet higher demand (peak load) periods. Small scale systems have long been used in such applications as propulsion of mine locomotives. Large scale applications must conserve the heat energy associated with compressing air; dissipating heat lowers the energy efficiency of the storage system.

Figure: *A compressed air locomotive used inside a mine between 1928 and 1961.*

The technology stores low cost off-peak energy, in the form of compressed air in an underground reservoir. The air is then released during peak load hours and, using older CAES technology, heated with the exhaust heat of a standard combustion turbine. This heated air is converted to energy through expansion turbines to produce electricity. A CAES plant has been in operation in McIntosh, Alabama since 1991 and has run successfully. Other applications are possible. Walker Architects published the first CO_2 gas CAES application, proposing the use of sequestered CO_2 for Energy Storage.

Compression of air creates heat; the air is warmer after compression. Expansion requires heat. If no extra heat is added, the air will be much colder after expansion. If the heat generated during compression can be stored and used during expansion, the efficiency of the storage improves considerably. There are three ways in which a CAES system can deal with the heat. Air storage can be adiabatic, diabatic, or isothermal. Several companies have also done design work for vehicles using compressed air power.

Flywheel Energy Storage: Flywheel energy storage (FES) works by accelerating a rotor (flywheel) to a very high speed and maintaining the energy in the system as rotational energy with the least friction losses possible. When energy is extracted from the system, the flywheel's rotational speed is reduced as a consequence of the principle of conservation of energy; adding energy to the system correspondingly results in an increase in the speed of the flywheel.

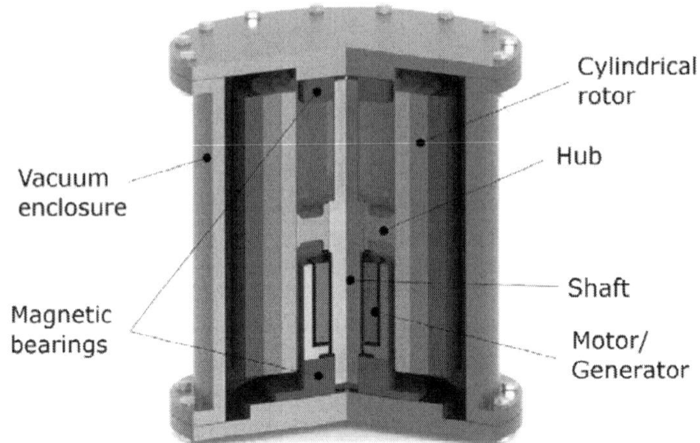

Figure: *The main components of a typical flywheel.*

Most FES systems use electricity to accelerate and decelerate the flywheel, but devices that directly use mechanical energy are being developed.

Advanced FES systems have rotors made of high strength carbon-fibre composites, suspended by magnetic bearings, and spinning at speeds from 20,000 to over 50,000 rpm in a vacuum enclosure. Such flywheels can come up to speed in a matter of minutes – reaching their energy capacity much more quickly than some other forms of storage.

A typical system consists of a rotor suspended by bearings inside a vacuum chamber to reduce friction, connected to a combination electric motor and electric generator.

Compared with other ways to store electricity, FES systems have long lifetimes (lasting decades with little or no maintenance; full-cycle lifetimes quoted for flywheels range from in excess of 10^5, up to 10^7, cycles of use), high energy density (100–130 W·h/kg, or 360–500 kJ/kg), and large maximum power output.

Gravitational Potential Energy Storage: A newer concept called potential energy storage or gravity energy storage system, has generated some proposals, at least one of which was under active development in 2013 in the U.S. state of Nevada in association with the California independent system operator. Whereas pumped hydro storage is a form of potential energy storage that uses water, the newer schemes are predicated on the movement of solid masses (such as hopper rail cars filled with plain earth driven by electric locomotives) from lower to higher elevations. The masses can then be stored there at a higher elevation with no loss of efficiency until power is required to be returned to the grid, at which time the masses are returned to their lower elevation storage site, generating electricity on their way down.

Advantages of one such system, called Advanced Rail Energy Storage (ARES), include the indefinite storage of potential energy with no efficiency losses over time (gravity doesn't degrade), the low bulk filler material costs when earth or rocks are used, the non-usage of water resources in areas where water is scarce, plus, since water is unused in that scheme there is no efficiency lost due to evaporation on hot days, one of several efficiency issues encountered with most pumped hydro storage reservoirs. As of 2014 ARES has started initial planning on a commercial-scale project in Nevada near its California border, partnered with Valley Electric Association Inc.

Thermal Storage: Thermal storage is the temporary storage or removal of heat for later use. An example of thermal storage is the storage of solar heat energy during the day to be used at a later time for heating at night. In the HVAC/R field, this type of application using thermal storage for heating is less common than using thermal storage for cooling. An example of the storage of "cold" heat removal for later use is ice made during the cooler night time hours for use during the hot daylight hours. This ice storage is produced when electrical utility rates are lower. This is often referred to as "off-peak" cooling.

When used for the proper application with the appropriate design, off-peak cooling systems can lower energy costs. The U.S. Green Building Council has developed the Leadership in Energy and Environmental Design (LEED) program to encourage the design of high-performance buildings that will help protect our environment. The increased levels of energy performance by utilising off-peak cooling may qualify of credits toward LEED Certification.

The advantages of thermal storage are:

- Commercial electrical rates are lower at night;
- it takes less energy to make ice when the ambient temperature is cool at night. Source energy (energy from the power plant) is saved.
- a smaller, less costly system can do the job of a much larger unit by running for more hours.

Ice Storage Air Conditioning: Air conditioning based on stored ice for thermal energy storage has become an accepted commercial technology in the 21st century. This is practical because of water's large heat of fusion: the melting of one metric ton of ice (approximately one cubic metre in size) can capture 334 megajoules (MJ) (317,000 BTU) of thermal energy.

Replacing existing air conditioning systems with ice storage based air conditioning offers a cost-effective energy storage method, enabling surplus wind energy and other such intermittent energy sources to be stored for use in chilling air at a later time, possibly months later. The most widely used form of this technology can be found in campus-wide air conditioning or chilled water systems of large buildings. Air conditioning systems, especially in commercial buildings, are the biggest contributors to peak electrical loads seen on hot summer days in various countries. In this application, a standard chiller runs at night to produce an ice pile. Water then circulates through the pile during the day to produce chilled water that would normally be the chiller's daytime output.

A partial storage system minimizes capital investment by running the chillers nearly 24 hours a day. At night, they produce ice for storage and during the day they chill water for the air conditioning system. Water circulating through the melting ice augments their production. Such a system usually runs in ice-making mode for 16 to 18 hours a day and in ice-melting mode for six hours a day. Capital expenditures are minimized because the chillers can be just 40 - 50% of the size needed for a conventional design. Ice storage sufficient to store half a day's rejected heat is usually adequate. A full storage system minimizes the cost of energy to run that system by entirely shutting off the chillers during peak load hours. The capital cost is higher, as such a system requires somewhat larger chillers than those from a partial storage system, and a larger ice storage system.

Electrochemical

Rechargeable Battery: A rechargeable battery, also called a storage battery or accumulator, is a type of electrical battery. It comprises one or more electrochemical cells, and is a type of energy accumulator. It is known as a 'secondary cell' because its electrochemical reactions are electrically reversible. Rechargeable batteries come in many different shapes and sizes, ranging from button cells to megawatt systems connected to stabilize an electrical distribution network. Several different combinations of chemicals are commonly used, including: lead–acid, nickel cadmium (NiCd), nickel metal hydride (NiMH), lithium ion (Li-ion), and lithium ion polymer (Li-ion polymer).

Figure: *A rechargeable battery bank used as an uninterruptible power supply in a data centree*

Rechargeable batteries have lower total cost of use and environmental impact than disposable batteries. Some rechargeable battery types are available in the same sizes as disposable types. Rechargeable batteries have higher initial cost but can be recharged very cheaply and used many times.

Common rechargeable battery chemistries include:

- Lead-acid battery: Lead acid batteries still hold the largest market share for all electric storage products today. A single lead-acid cell produces about 2V when charged. In the charged state the metallic lead negative electrode and the lead sulfate positive electrode are immersed in a dilute sulfuric acid (H2SO4) electrolyte. In the discharge process electrons are pushed out of the cell as lead sulfate is formed at the negative electrode while the electrolyte is reduced to water.
- Nickel–cadmium battery (NiCd): Uses nickel oxide hydroxide and metallic cadmium as electrodes. Cadmium is a toxic element, and was banned for most uses by the European Union in 2004. Nickel–cadmium batteries have been almost completely superseded by nickel–metal hydride (NiMH) batteries.
- Nickel–metal hydride battery (NiMH): First commercial types were available in 1989. These are now a common consumer and industrial type. The battery has a hydrogen-absorbing alloy for the negative electrode instead of cadmium.
- Lithium-ion battery: The technology behind the lithium-ion battery has not yet fully reached maturity. However, the batteries are the type of choice in many consumer electronics and have one of the best energy-to-mass ratios and a very slow loss of charge when not in use.
- Lithium-ion polymer battery: These batteries are light in weight and can be made in any shape desired.

Flow Battery: A flow battery is a type of rechargeable battery where rechargeability is provided by two chemical components dissolved in liquids contained within the system and separated by a membrane. Ion exchange (providing flow of electrical current) occurs through the membrane while both liquids circulate in their own respective space. Cell voltage is chemically determined by the Nernst equation and ranges, in practical applications, from 1.0 to 2.2 Volts.

A flow battery is technically akin both to a fuel cell and an electrochemical accumulator cell (electrochemical reversibility). While it has technical advantages such as potentially separable liquid tanks and near unlimited longevity over most conventional rechargeables, current implementations are comparatively less powerful and require more sophisticated electronics. Newer types of flow batteries are being developed to provide bulk energy storage, since increasing a system's overall energy capacity (measured in MWh) basically requires only an increase in the size of its liquid chemical storage reservoirs.

Supercapacitors: Supercapacitors, also called electric double-layer capacitors (EDLC) or ultracapacitors, are generic terms for a family of electrochemical capacitors. Supercapacitors don't have conventional solid dielectrics. The capacitance value of an electrochemical capacitor is determined by two storage principles, which both contribute indivisibly to the total capacitance:

Supercapacitors bridge the gap between conventional capacitors and rechargeable batteries. They store the most energy per unit volume or mass (energy density) among capacitors. They support up to 10,000 farads/1.2 volt, up to 10,000 times that of electrolytic capacitors, but deliver or accept less than half as much power per unit time (power density).

Figure: *One of a fleet of electric capabuses powered by supercapacitors, at a quick-charge station-bus stop, in service during Expo 2010 Shanghai China. Charging rails can be seen suspended over the bus.*

By contrast, while supercapacitors have energy densities that are approximately 10% of conventional batteries, their power density is generally 10 to 100 times greater. This results in much shorter charge/discharge cycles than batteries. Additionally, they will tolerate many more charge and discharge cycles than batteries.

Supercapacitors support a broad spectrum of applications, including:

- Low supply current for memory backup in static random-access memory (SRAM)
- Power for cars, buses, trains, cranes and elevators, including energy recovery from braking, short-term energy storage and burst-mode power delivery

UltraBattery: The UltraBattery is a hybrid lead-acid cell and carbon-based ultracapacitor (or supercapacitor) invented by Australia's national research body, the Commonwealth Scientific and Industrial Research Organisation (CSIRO). The lead-acid cell and ultracapacitor share the sulfuric acid electrolyte and both are packaged into the same physical cell. The UltraBattery can be manufactured with similar physical and electrical characteristics to conventional lead-acid batteries making it possible to cost-effectively replace many existing lead-acid installations with UltraBattery technology.

The key difference between conventional lead-acid batteries and UltraBattery technology is that the UltraBattery performs like an ultracapacitor when necessary and like a lead-acid cell at other times meaning it can work across a very wide range of applications. The constant cycling and fast charging and discharging necessary for applications such as renewable smoothing, grid regulation, electric and hybrid-electric vehicles can have deleterious effects on chemical batteries but are well handled by the ultracapacitive qualities of UltraBattery technology.

The UltraBattery will tolerate high charge and discharge levels and very large numbers of cycles during its lifetime, outperforming previous lead-acid cells by more than an order of magnitude. In hybrid-electric vehicle tests, millions of cycles have been achieved. The UltraBattery is also highly tolerant to the effects of sulfation compared with traditional lead-acid cells. This means it can operate continuously in partial state of charge whereas traditional lead-acid batteries are generally held at full charge between discharge events. It is generally electrically inefficient to fully charge a lead-acid battery so by decreasing time spent in the top region of charge the UltraBattery achieves high efficiencies, typically between 85 to 95% DC-DC.

The technology has been installed in Australia and the U.S.A. on the megawatt scale performing frequency regulation and renewable smoothing applications.

Other Chemical

Hydrogen: Hydrogen is also being developed as an electrical power storage medium. Hydrogen is not a primary energy source, but a portable energy storage method, because it must first be manufactured by other energy sources in order to be used. However, as a storage medium, it may be a significant factor in using renewable energies.

With intermittent renewables such as solar and wind, the output may be fed directly into an electricity grid. At penetrations below 20%

of the grid demand, this does not severely change the economics; but beyond about 20% of the total demand, external storage will become important. If these sources are used for electricity to make hydrogen, then they can be utilised fully whenever they are available, opportunistically. Broadly speaking, it does not matter when they cut in or out, the hydrogen is simply stored and used as required. A community based pilot program using wind turbines and hydrogen generators is being undertaken from 2007 for five years in the remote community of Ramea, Newfoundland and Labrador. A similar project has been going on since 2004 on Utsira, a small Norwegian island municipality.

Energy losses are involved in the hydrogen storage cycle of hydrogen production for vehicle applications with electrolysis of water, liquification or compression, and conversion back to electricity. and the hydrogen storage cycle of production for the stationary fuel cell applications like MicroCHP at 93 % with biohydrogen or biological hydrogen production, and conversion to electricity.

About 50 kW·h (180 MJ) of solar energy is required to produce a kilogram of hydrogen, so the cost of the electricity clearly is crucial, even for hydrogen uses other than storage for electrical generation. At $0.03/kWh, common off-peak high-voltage line rate in the United States, this means hydrogen costs $1.50 a kilogram for the electricity, equivalent to $1.50 a U.S. gallon for gasoline if used in a fuel cell vehicle. Other costs would include the electrolyzer plant, hydrogen compressors or liquefaction, storage and transportation, which will be significant.

Underground Hydrogen Storage: Underground hydrogen storage is the practice of hydrogen storage in underground caverns, salt domes and depleted oil and gas fields. Large quantities of gaseous hydrogen have been stored in underground caverns by Imperial Chemical Industries (ICI) for many years without any difficulties. The European project Hyunder indicated in 2013 that for the storage of wind and solar energy, an additional 85 caverns are required as it can't be covered by PHES and CAES systems.

Power to Gas: Power to gas is a technology which converts electrical power to a gaseous fuel. There are currently three methods in use; all use electricity to split water into hydrogen and oxygen by means of electrolysis.

In the first method, the resulting hydrogen is injected into the natural gas grid or is used in transport or industry. The second method is to combine the hydrogen with carbon dioxide and convert the two gases to methane using a methanation reaction such as the Sabatier reaction, or biological methanation resulting in an extra energy conversion loss of 8%.

The methane may then be fed into the natural gas grid. The third method uses the output gas of a wood gas generator or a biogas plant, after the biogas upgrader is mixed with the produced hydrogen from the electrolyzer, to upgrade the quality of the biogas.

The excess power or off–peak power generated by wind turbines or solar arrays can then be used for load balancing in an energy grid. Using the existing natural gas system for hydrogen, fuel cell manufacturer Hydrogenics and natural gas distributor Enbridge have teamed up to develop such a power to gas system in Canada.

Hydrogen can be stored in natural gas pipeline networks. Before switching to natural gas, German gas networks were operated using towngas, which for the most part consisted of hydrogen. The storage capacity of the German natural gas network, which also contains many man-made caverns (artificial caves produced by mining), is more than 200,000 GW·h, which is enough for several months of energy requirement. By comparison, the capacity of all German pumped storage power plants amounts to only about 40 GW·h. The transport of energy through a gas network is done with much less loss (<0.1%) than in a power network (8%) (besides High-voltage direct current). The use of the existing natural gas pipelines for hydrogen was studied by NaturalHy.

Biofuels: Various biofuels such as biodiesel, straight vegetable oil, alcohol fuels, or biomass can be used to replace hydrocarbon fuels. Various chemical processes can convert the carbon and hydrogen in coal, natural gas, plant and animal biomass, and organic wastes into short hydrocarbons suitable as replacements for existing hydrocarbon fuels. Examples are Fischer-Tropsch diesel, methanol, dimethyl ether, or syngas. This diesel source was used extensively in World War II in Germany, with limited access to crude oil supplies. Today South Africa produces most of the country's diesel from coal for similar reasons. A long term oil price above US$35/bbl may make such synthetic liquid fuels economical on a large scale. Some of the energy in the original source is lost in the conversion process. Historically, coal itself has been used directly for transportation purposes in vehicles and boats using steam engines. Additionally, compressed natural gas is also used as fuel, for instance for buses with some mass transit agencies.

Methane: Methane is the simplest hydrocarbon with the molecular formula CH_4. Methane can be produced from electricity using power to gas technologies. Methane is more easily stored than hydrogen and the transportation, storage and combustion infrastructure (pipelines, gasometers, power plants) are mature.

Synthetic natural gas (SNG) can be created in a multi-step process, starting when hydrogen and oxygen are produced during the electrolysis of water. Hydrogen would then be reacted with carbon dioxide in a Sabatier process, producing methane and water. Methane can be stored and used to produce electricity later. The water produced would be recycled back to the electrolysis stage, reducing the need for additional new pure water. In the electrolysis stage oxygen would also be stored for methane combustion in a pure oxygen environment at an adjacent power plant, eliminating nitrogen oxides.

In the combustion of methane, carbon dioxide (CO_2) and water are produced. The carbon dioxide created would be recycled back to boost the Sabatier process and water would be recycled back to the electrolysis stage. The carbon dioxide produced by methane combustion would be turned back to methane, thus producing no greenhouse gases. Methane production, storage and adjacent combustion would recycle all the reaction products, creating a low carbon cycle.

The CO_2 would therefore be a resource having economic value as a component of an energy storage vector, not a cost as in Carbon capture and storage.

Aluminium, Boron, Silicon, and Zinc

Aluminium, Boron, silicon, lithium, and zinc have been proposed as energy storage solutions.

Electrical Methods

Capacitor: A capacitor (originally known as a 'condenser') is a passive two-terminal electrical component used to store energy electrostatically in an electric field. The forms of practical capacitors vary widely, but all contain at least two electrical conductors (plates) separated by a dielectric (i.e., insulator). A capacitor can store electric energy when disconnected from its charging circuit, so it can be used like a temporary battery, or like other types of rechargeable energy storage system. Capacitors are also commonly used in electronic devices to maintain power supply while batteries are being changed. (This prevents loss of information in volatile memory.) Conventional capacitors provide less than 360 joules per kilogram of energy density, whereas a conventional alkaline battery has a density of 590 kJ/kg. Unlike a resistor, a capacitor does not dissipate energy. Instead, a capacitor stores energy in the form of an electrostatic field between its plates. When there is a potential difference across the conductors (e.g., when a capacitor is attached across a battery), an electric field develops across the dielectric, causing positive charge (+Q) to collect on one plate and negative charge (-Q) to collect on the other plate. If a battery has been attached to a capacitor for a sufficient amount of time, no current can flow through the capacitor. However, if an accelerating or alternating voltage is applied across the leads of the capacitor, a displacement current can flow.

The capacitance is greater when there is a narrower separation between conductors and when the conductors have a larger surface area. In practice, the dielectric between the plates passes a small amount of leakage current and also has an electric field strength limit, known as the breakdown voltage. The conductors and leads introduce an undesired inductance and resistance. Capacitors are widely used in electronic circuits for blocking direct current while allowing alternating current to pass. In analog filter networks, they smooth the output of power supplies. In resonant circuits they tune radios to particular frequencies. In electric power transmission systems they stabilize voltage and power flow.

Electromagnetic Storage

Superconducting Magnetic Energy Storage (SMES) systems store energy in a magnetic field created by the flow of direct current in a superconducting coil which has been cryogenically cooled to a temperature below its superconducting critical temperature. A typical SMES system includes three parts: superconducting coil, power conditioning system and cryogenically cooled refrigerator. Once the superconducting coil is charged, the current will not decay and the magnetic energy can be stored indefinitely.

The stored energy can be released back to the network by discharging the coil. The power conditioning system uses an inverter/rectifier to transform alternating current (AC) power to direct current or convert DC back to AC power. The inverter/rectifier accounts for about 2–3% energy loss in each direction. SMES loses the least amount of electricity in the energy storage process compared to other methods of storing energy. SMES systems are highly efficient; the round-trip efficiency is greater than 95%.

Due to the energy requirements of refrigeration and the high cost of superconducting wire, SMES is currently used for short duration energy storage. Therefore, SMES is most commonly devoted to improving power quality. If SMES were to be used for utilities it would be a diurnal storage device, charged from baseload power at night and meeting peak loads during the day.

Battery (Electricity)

An electric battery is a device consisting of one or more electrochemical cells that convert stored chemical energy into electrical energy. Each cell contains a positive terminal, or cathode, and a negative terminal, or anode. Electrolytes allow ions to move between the electrodes and terminals, which allows current to flow out of the battery to perform work.

Primary (single-use or "disposable") batteries are used once and discarded; the electrode materials are irreversibly changed during discharge. Common examples are the alkaline battery used for flashlights and a multitude of portable devices. Secondary (rechargeable batteries) can be discharged and recharged multiple times; the original composition of the electrodes can be restored by reverse current. Examples include the lead-acid batteries used in vehicles and lithium ion batteries used for portable electronics.

Batteries come in many shapes and sizes, from miniature cells used to power hearing aids and wristwatches to battery banks the size of rooms that provide standby power for telephone exchanges and computer data centrees.

According to a 2005 estimate, the worldwide battery industry generates US$48 billion in sales each year, with 6% annual growth.

Batteries have much lower specific energy (energy per unit mass) than common fuels such as gasoline. This is somewhat mitigated by the fact that batteries deliver their energy as electricity (which can be converted efficiently to mechanical work), whereas using fuels in engines entails a low efficiency of conversion to work.

Principle of Operation

Batteries convert chemical energy directly to electrical energy. A battery consists of some number of voltaic cells. Each cell consists of two half-cells connected in series by a conductive electrolyte containing anions and cations. One half-cell includes electrolyte and the negative electrode, the electrode to which anions (negatively charged ions) migrate; the other half-cell includes electrolyte and the positive electrode to which cations (positively charged ions) migrate. Redox reactions power the battery. Cations are reduced (electrons are added) at the cathode during charging, while anions are oxidized (electrons are removed) at the anode during discharge. The electrodes do not touch each other, but are electrically connected by the electrolyte. Some cells use different electrolytes for each half-cell. A separator allows ions to flow between half-cells, but prevents mixing of the electrolytes.

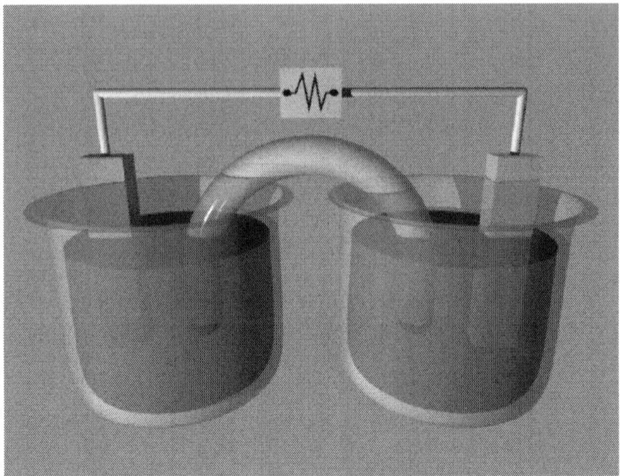

Figure: *A voltaic cell for demonstration purposes. In this example the two half-cells are linked by a salt bridge separator that permits the transfer of ions.*

Each half-cell has an electromotive force (or emf), determined by its ability to drive electric current from the interior to the exterior of the cell. The net emf of the cell is the difference between the emfs of its half-cells. Thus, if the electrodes have emfs $\mathcal{E}_1$ and $\mathcal{E}_2$, then the net emf is $\mathcal{E}_2 - \mathcal{E}_1$; in other words, the net emf is the difference between the reduction potentials of the half-reactions.

The electrical driving force or ΔV_{bat} across the terminals of a cell is known as the *terminal voltage (difference)* and is measured in volts. The terminal voltage of a cell that is neither charging nor discharging is called the open-circuit voltage and equals the emf of the cell. Because of internal resistance, the terminal voltage of a cell that is discharging is smaller in magnitude than the open-circuit voltage and the terminal voltage of a cell that is charging exceeds the open-circuit voltage.

An ideal cell has negligible internal resistance, so it would maintain a constant terminal voltage of ε until exhausted, then dropping to zero. If such a cell maintained 1.5 volts and stored a charge of one coulomb then on complete discharge it would perform 1.5 joules of work. In actual cells, the internal resistance increases under discharge and the open circuit voltage also decreases under discharge. If the voltage and resistance are plotted against time, the resulting graphs typically are a curve; the shape of the curve varies according to the chemistry and internal arrangement employed.

The voltage developed across a cell's terminals depends on the energy release of the chemical reactions of its electrodes and electrolyte. Alkaline and zinc–carbon cells have different chemistries, but approximately the same emf of 1.5 volts; likewise NiCd and NiMH cells have different chemistries, but approximately the same emf of 1.2 volts. The high electrochemical potential changes in the reactions of lithium compounds give lithium cells emfs of 3 volts or more.

Fuel Cell

A fuel cell turns hydrogen, the fuel, into electricity using air and other catalysts. The fuel cell harnesses chemical energy trapped in hydrogen gas and converts it into the kinetic energy we know as electricity, without fossil fuels, combustion, or polluting emissions. As a remarkably efficient, incredibly clean source of renewable energy, fuel cells can take the place of both batteries and engines to power vehicles, laptops, and residential power grids.

The technology of fuel cells has been well researched and developed, although for economic and political reasons, they have not been widely implemented. Such a clean power source, as part of a widespread hydrogen economy, guarantees less dependence on the dwindling supplies of fossil fuels, creates less greenhouse gases that contribute to global climate change, and does not explode or malfunction as frequently as engine-driven electricity.

Chemically, a fuel cell takes in hydrogen and air, creates electricity, and produces byproducts of water and heat. The outside layers of a fuel cell are the anode plate, with a positive charge, and the cathode plate, with a negative charge. Together with the centree electrolyte plate, they are catalytic environments that encourage certain electrochemical functions.

The anode separates hydrogen into protons and electrons. The electrons flow along a path, producing electrical current for a circuit, while protons move through the electrolyte to the cathode. The cathode combines oxygen with protons, as well as collecting some of the electrons in the circuit, to recombine them into water. The cathode allows water, and extra heat, to be used as additional sources of energy.

An independent hydrogen supply, as from a tank at a station, doesn't need to be the sole source of fuel in a fuel cell. In fact, one can run on rotting organic material, like vegetation, because that gives off hydrogen, too. Or hydrogen might be separated from oxygen out of water, through electrolysis, by solar or wind power. If water is used as a source of hydrogen, the fuel cell is practically immortal, as it continues the cycle from water to hydrogen to water.

Additionally, a fuel cell is flexible because it can be small and portable or larger and permanent. The cost to convert to a fuel cell-driven power grid may initially be high, but over time it will significantly reduce the costs of maintenance, repair, and fuel compared to conventional electricity generators. If one can use the resulting heat, say, to warm a house during winter, the fuel cell becomes even more cost effective.

Bibliography

Andrew, N.: *Foundations of Nanomechanics: from Solid-state Theory to Device Applications*, Berlin, New York, 2003.

Attebery, B.: *Nanoculture: Implications of the New Technoscience*, Intellect Books, Bristol, 2004.

Biggar, John: *The Andes: A Guide for Climbers,* Andes Publishing, Scotland, 2005.

Braun, Bruce; Castree, Noel: *Remaking Reality: Nature at the Millenium*, Routledge, UK., 1998.

Brosnan, J.: *The Primal Scream: A History of Science Fiction Film*, Orbit Books, London, 1991.

Caneva, K. L.: *Robert Mayer and the Conservation of Energy*, Princeton University Press, Princeton, New Jersey, 1993.

Carroll, J.S.: *Social Science Research Methods for Assessing Societal Implications of Nanotechnology*, Dordrecht, Kluwer, 2001.

Clegg, Brian: *Eco-logic: Cutting Through the Greenwash: Truth, Lies and Saving the Planet,* Eden Project, London, 2009.

Crow, M. & Sarewitz, D.: *Societal Implications of Nanoscience and Nanotechnology*, Dordrecht: Kluwer, 2001.

Daily, Gretchen C.: *Nature's Services: Societal Dependence on Natural Ecosystems*, Island Press, Washington, D.C., 1997.

Dobzhanshy, T.: *Genetics of the Evolutionary Process*, Columbia University Press, New York, 1970.

Dunai, T.J.: *Cosmogenic Nucleides,* Cambridge University Press, U.K., 2010.

Dyson, Tim: *India's Historical Demography: Studies in Famine, Disease, and Society*, Riverdale, Westwood, MA, 1989.

Eamonn Gearon: *The Sahara: A Cultural History*, Signal Books, UK, 2011.

Engelmann, S., & Carnine, D.: *Theory of Instruction: Principles and Applications*, ADI Press, Eugene, OR, 1991.

Feynman, Richard: *The Feynman Lectures on Physics;* Addison Wesley, U.S.A., 1964.

Gibson, Donald: *Environmentalism: Ideology and Power*, Nova Science Publication, New York, 2003.

Gould, Peter C.: *Early Green Politics*, Harvester Press, Brighton, 1988.

Haber, L. F.: *The Chemical Industry: 1900-1930, International Growth and Technological Change*, Oxford: Claredon Press, 1971.

Hamilton, Adrian: *Oil: The Price of Power*. London: Michael Joseph/Rainbird, 1986.

Jansson, AnnMari: *Investing in Natural Capital : The Ecological Economics Approach to Sustainability*, Island Press, Washington, D.C.1994.

Johnson, Chris: *Australia's Mammal Extinctions*, Cambridge University Press, Melbourne, 2006.

Krzeminska, Joanna: *Are Support Schemes for Renewable Energies Compatible with Competition Objectives? An Assessment of National and Community Rules*, Oxford University Press, London, 2007

Lippert, I.: *Greenwashing*, Sage Publications, New Delhi, 2011.

Mach, E.: *History and Root of the Principles of the Conservation of Energy*. Open Court Pub. Co., IL, 1872.

Mayr, E.: *Change of Genetic Environment and Evolution*, Allen and Unwin, London, 1954.

Nash, L. K.: *Elements of Statistical Thermodynamics*, Addison-Wesley, Reading, MA, 1974.

Odell, Peter R.: *Oil and World Power: Background of the Oil Crisis,* New York: Viking Penguin, 1986.

Pearce, D.: *Measuring Sustainable Development*, Earthscan, United States, 1993.

Pressman, J.: *Nano Narrative: A Parable from Electronic Literature,* Intellect Books, Bristol, 2004.

Pressman, J.: *Nanoculture: Implications of the New Technoscience*, Intellect Books, Bristol, UK, 2004.

Rothwell, Edward J.; Cloud, Michael J.: *Electromagnetics*, CRC Press, US, 2001.

Shanefield, D. J.: *Industrial Electronics for Engineers, Chemists, and Technicians,* William Andrew Publishing, Norwich, NY, 2001.

Sutton, A.P.: *Electronic Structure of Materials*. Oxford Science Publications, 1993.

Tarkin, Laurie: *Electromagnetic Fields: What You Need to Know to Protect Your Health*, Bantam, 1994.

Tipler, Paul: *Physics for Scientists and Engineers, Light, Electricity and Magnetism,* W. H. Freeman, New York, 1998.

Wangsness, Roald K.: *Electromagnetic Fields,* Wiley, New York, 1986.

Wayne, H. Beaty: *Standard Handbook for Electrical Engineers,* McGraw-Hill, New York, 1978.

Index

❑❑❑